ENVIRONMENT AND MAN
VOLUME FIVE

The Marine Environment

General Editors

John Lenihan
O.B.E., M.Sc., Ph.D., C.Eng., F.I.E.E., F.Inst.P., F.R.S.E.

Director of the Department of Clinical Physics and Bio-Engineering, West of Scotland Health Boards; Professor of Clinical Physics, University of Glasgow; Chairman of the Scottish Technical Education Council.

and

William W Fletcher
B.Sc., Ph.D., F.L.S., F.I.Biol., F.R.S.E.

Professor of Biology and Past Dean of the School of Biological Sciences, University of Strathclyde; Chairman of the Scottish Branch of the Institute of Biology; President of the Botanical Society of Edinburgh.

Blackie
Glasgow and London

Blackie & Son Limited
Bishopbriggs
Glasgow G64 2NZ

©1977 Blackie & Son Ltd.
First published 1977

Acc. No.	32087
Class No.	574.592
Date Rec	9 JUN 1981
Order No.	F25783

All rights reserved
No part of this publication may be reproduced,
stored in a retrieval system, or transmitted,
in any form or by any means,
electronic, mechanical, recording or otherwise,
without prior permission of the Publishers

International Standard Book Numbers

Paperback 0 216 90130 8

Hardback 0 216 90131 6

Printed in Great Britain by
Thomson Litho Ltd., East Kilbride, Scotland

Background to Authors

Environment and Man: Volume Five

PAUL TETT, M.A., Ph.D., is a Principal Scientific Officer at the Dunstaffnage Marine Research Laboratory of the Scottish Marine Biological Association. From 1974 to 1975 he was Research Associate Professor in the Department of Environmental Sciences, University of Virginia.

ANTHONY NELSON-SMITH, B.Sc., Ph.D., F.Z.S., is Senior Lecturer in Zoology at the University College of Swansea. For the last sixteen years he has studied the effects of various pollutants on the ecology of seashores and estuaries, and helped to set up the Oil Pollution Research Unit at Orielton, Pembroke in 1967.

ERIC JOHN PERKINS, B.Sc., Ph.D., F.L.S., is Senior Lecturer in Biology at the University of Strathclyde, and Head of the University's Marine Laboratory at Kilcreggan. Prior to his present appointment, he was employed by the Atomic Energy Authority, the Central Electricity Generating Board, and the Department of Agriculture and Fisheries for Scotland. In 1969 he was awarded a Winston Churchill Travelling Fellowship to study effluent problems in North America.

DAVID IAN HUNTER BARR, B.Sc., Ph.D., F.I.C.E., M.A.S.C.E., is Professor of Hydraulics in the Department of Civil Engineering, University of Strathclyde. After industrial experience in planning and design aspects of electricity generation, he returned to Strathclyde University as a research student to study problems of heat dissipation from thermal power stations. He later joined the staff and has maintained interest in problems relating to power generation.

ROBERT SIMPSON SILVER, C.B.E., M.A., D.Sc., F.Inst.P., F.I.Mech.E., F.R.S.E., is James Watt Professor of Mechanical Engineering at The University of Glasgow. He has been associated with The Weir Group Limited since 1939, and is presently a Director of Weir Westgarth and Weir Heat Exchange Limited. In 1968 he was awarded the UNESCO Science Prize for fundamental scientific work and its useful application in developing countries.

WILLIAM SCHOOLBRAID McCARTNEY, B.Sc., Grad.M.I.Mech.E., is Lecturer in Mechanical Engineering at the University of Glasgow. His research involves equilibration studies and their application to desalination.

Series Foreword

MAN IS A DISCOVERING ANIMAL—SCIENCE IN THE SEVENTEENTH CENTURY, scenery in the nineteenth and now the environment. In the heyday of Victorian technology—indeed until quite recently—the environment was seen as a boundless cornucopia, to be enjoyed, plundered and re-arranged for profit.

Today many thoughtful people see the environment as a limited resource, with conservation as the influence restraining consumption. Some go further, foretelling large-scale starvation and pollution unless we turn back the clock and adopt a simpler way of life.

Extreme views—whether exuberant or gloomy—are more easily propagated, but the middle way, based on reason rather than emotion, is a better guide for future action. This series of books presents an authoritative explanation and discussion of a wide range of problems related to the environment, at a level suitable for practitioners and students in science, engineering, medicine, administration and planning. For the increasing numbers of teachers and students involved in degree and diploma courses in environmental science the series should be particularly useful, and for members of the general public willing to make a modest intellectual effort, it will be found to present a thoroughly readable account of the problems underlying the interactions between man and his environment.

Preface

THE SEA IS A VAST ENVIRONMENTAL LABORATORY FROM WHICH WE HAVE much to learn. The marine environment is an ecological system characterized by great stability, and is indeed virtually unchanged by man's efforts at rearrangement.

The constancy of the marine environment does not denote unwillingness on the part of man to upset it, for the sea is used increasingly as a rubbish dump, a water-tank, a foodstore and a source of power. But the inherent stability of the system is so great that these insults produce only minor perturbations, although local or temporary effects may appear more substantial.

The sea is important to students of the environment as a potential source of great wealth which has not yet been damaged significantly by the carelessness, greed or stupidity of man. Production and pollution, the main issues in managing it, are the essential themes of this volume.

It is often claimed that the world's protein supply could be greatly augmented by fish farming and other techniques aimed at increasing the harvest of the sea. Dr Paul Tett, in his study of marine production, explains that man has made little progress in managing the ecology of the sea; we are still predators rather than farmers of fish.

The task of increasing the supply is not easy. The global production of fish suitable for human food is about 100 million tons a year, and more than half of this is already being caught. Efforts to increase the catch will need careful management if temporary gains are not to be nullified by the long-term consequences of over-fishing. The present yield of 50–60 million tons per year might eventually be doubled by exploiting deep-water fish, squids and crustaceans, but new types of gear and new methods of fishing will have to be developed. In any event, much of the new yield would probably be made into fish meal and ultimately eaten after conversion to chicken or pork.

Fish farming is unlikely to be economically viable in Western countries but offers attractive possibilities in the Third World. Dr Tett concludes by reminding us that the coastal waters which are so important to commercial fisheries (and to fish farming) are vulnerable to the harmful effects of industry and other human activity.

Pollution of beaches by oil spillage is an environmental hazard that has generated much anxiety in recent times. Dr A. Nelson-Smith shows how the problem has grown during the past 50 years, through the necessity of transporting great amounts of oil from the producing to the consuming countries. Off-shore oil production is a new source of potential trouble. The amount of oil entering the sea every year is now about 2 million tons. Only 5% of this comes from collisions and other spectacular accidents; much larger contributions come from the routine operation of tankers and other ships, refineries and, indirectly, from industrial and motor-car oil wastes.

Dr Nelson-Smith gives a detailed account of the harmful effects of oil pollution on marine birds, animals and plants; fish are able on the whole to avoid the consequences of all but the most disastrous spills; however, oil contains many carcinogenic substances which may find their way into food chains and constitute a hazard to human health. In discussing the techniques available for the treatment of oil spills, he recalls that most of the damage which followed the stranding of the *Torrey Canyon* off the Cornish coast resulted from the ill-judged application of detergent mixtures which were more toxic than the original oil.

Dr E. J. Perkins assesses a complementary problem in his examination of the ecological effects associated with the dumping of inorganic wastes in coastal waters. After a general review of the problem and the experimental techniques which have been developed to deal with it, he presents some illuminating case histories, including two investigations in which careful study showed that large-scale dumping of apparently toxic material produced no adverse ecological effects. These studies illustrate the ability of the living environment to adapt to change, even of a wholly unnatural kind. Dr. Perkins underlines some popular errors and misunderstandings which can lead the inexperienced investigator to erroneous conclusions. The significant message of his chapter is that the wealth of technology now available for measuring and monitoring the chemical environment must be supported by sound scientific insight, if it is to be useful in dealing with pollution problems.

The tides convert great amounts of gravitational energy into low-grade heat—a process which clearly calls for the intervention of the engineer. Professor D. I. H. Barr observes that tidal mills, diverting some of the gravitational energy for useful purposes, were successfully developed centuries ago. A few survived into modern times, but the cheaper sources of energy available after the Industrial Revolution became more popular. Recent advances in theory and practice have made the harnessing of the tides technically feasible on a considerable scale. After thorough assessment of the engineering and economic aspects of the problem, Professor Barr finds that tidal power is not yet economically viable, although future changes in energy policy may alter the situation. He concludes his chapter by reviewing recent work on the extraction of energy from ocean waves—an exciting prospect, but one which is not likely to achieve practical fulfilment for another two decades.

The most obvious use of the ocean is as a water supply. The sea is the original abode of life, providing our primitive ancestors with a stable climate, adequate nutrition, simple recycling of waste and virtually zero gravity. But life has since developed in such a way that terrestrial creatures need the solvent rather than the solution. The sun's energy distils seawater very successfully, delivering it as rain, but rather unevenly, with the result that in many parts of the world the demand for water exceeds the supply.

In their survey of desalination, Professor R. S. Silver and Mr W. S. McCartney show that the solar still can be augmented at a cost which is already attractive in regions of insufficient rainfall and is likely to become attractive even in developed countries, where the growth of industry and other changes in society create demands that cannot be met from conventional water resources. They show that agriculture needs about 10 times as much water as manufacturing industry, per ton of product. Consequently the deserts in arid but energy-rich countries may be covered, not by crops, but by factories making goods which can be exported to buy food.

The oceans cover most of the earth, and it is not to be expected that every important aspect of the marine environment can be examined in a single volume. The role of the oceans in stabilizing climate has been referred to in earlier volumes but calls for fuller treatment later in this series. The part played by ocean water and sediments in the chemical conversion and recycling of many toxic elements is a matter which will be discussed in a future volume. The possible exploitation of the mineral wealth in the sea (as distinct from the hydrocarbon resources under the sea) represents a further substantial topic for future study.

Meanwhile this volume is offered as an objective survey of some of the benefits that man can enjoy, and some of the hazards that he must face, in a responsible approach to the marine environment.

JOHN LENIHAN
WILLIAM W. FLETCHER

Contents

CHAPTER ONE— **MARINE PRODUCTION**   1
by Paul Tett

Introduction. The natural history of the sea. Life in the sea. Food chains and nutrient cycling. **Marine production ecology.** Primary production. Photosynthesis. Standing crop. Seasonal and regional variations in primary production. Global production. Production at higher trophic levels. Ecological efficiency and potential fish production. **The fisheries.** Introduction. The herring. The plaice. Prudent predators and efficient prey. Recruitment. Problems of the fisheries. **Future exploitation of marine production.** The fisheries. Mariculture. Conclusion. Further reading.

CHAPTER TWO— **BIOLOGICAL CONSEQUENCES OF OIL SPILLS** by A. Nelson-Smith   46

Causes of marine oil pollution. The development and nature of the oil industry. The possibility of spillage. The discharge of ballast. The amount of oil reaching the sea. **Properties of spilt oil.** Classification of oil pollutants. Components of crude oil. Degradation of oil components. **Biological effects.** External. Internal. **Acute toxicity of oils.** Median lethal concentration. Range of sensitivity. Results of toxicity tests. **Sublethal effects.** Plankton. Eggs and larvae. Effects on behaviour. Concentration of chemicals by oil. **Wider implications. Ecological consequences.** Disturbance of the ecosystem and succession. Human amenity. **Carcinogenesis. Treatment of oil spills.** The need to be prepared. Methods of removal. **Further reading.**

CHAPTER THREE— **INORGANIC WASTES**   70
by E. J. Perkins

Introduction. The continental shelf. The inorganic wastes. **The problem in general.** Variations in pH. pH and toxicity. The dangers of using data from fresh-water situations. **The problem of dumping.** Iron waste deposits. Sludge deposits. Trace elements and *Nereis*. China clay. **The Solway Firth.** Slag disposal. Chemical content. The scarcity of the lugworm. **Phosphogypsum.** Survey results. Concentration of sulphate. Interpretation of data. **Eutrophication.** "Red tides". Green algae. Sewage treatment. A hypothesis. **Commentary. Further reading.**

CHAPTER FOUR— **POWER FROM THE TIDES AND WAVES**   102
by D. I. H. Barr

Historical and introductory sketch. Early tide mills. The Industrial Revolution brings cheaper power. Long-term studies leading to modern concepts. Technical considerations. **The oceanic and nearshore tides.** The idealized concept of the equilibrium tides. A basic astronomical relationship with tidal range. Tidal conditions for potential power projects. **Turbine and pump units.** Principle of change of angular momentum of water flow. Constant-speed running for alternating current. Requirement of submergence to prevent cavitation. Recent turbine developments. Bulb turbines. Peripheral-generator straight-flow turbines. S-tube turbines. **Other components.** Alternators, dynamos or other energy conversion methods. Sluices. Active barrage framework and float-in proposals. Non-active barrage construction. **Methods of operation and integration with power distribution system.** Single-basin uni-directional power production. Single-basin two-directional power production. Double-basin uni-direction power production. Other basin arrangements. Pumping to augment output, and matching of pumped storage with tidal power production. **Installations and projects.** La Ranche estuary. Kislaya Guba experimental tidal-power plant. Severn Estuary and Bristol Channel. Bay of Fundy. Passamaquoddy Bay. Head of Bay projects. Cook Inlet. San José Gulf. Other sites and more fundamental developments. **Assessment. Power from waves.** Introductory. Historical. The Salter shape—the "nodding duck" system. Other systems and comparative testing programme. The ocean wave climate. The problem conditions and their possible economic consequences. Prospects for power from the waves. **Further reading.**

CHAPTER FIVE— **DESALINATION**   138
by R. S. Silver and W. S. McCartney

Introduction. Desalination and the environment. **Water supply and desalination in an industrial society.** Water, the forgotten industry. Water and energy. Energy requirements of desalination. The cost of water by desalination. Desalination and agriculture. **Methods of desalination.** The distillation processes. Reverse osmosis. The freezing process. Distillation and power installation. Chemical aspects of distillation. Effects of desalination plant effluents on the marine environment. Appearance and amenity aspects of desalination plant. **Concluding survey. Further reading.**

INDEX   167

CHAPTER ONE

MARINE PRODUCTION

PAUL TETT

Introduction

The fifty-two million tons of marine fish, molluscs and crustaceans landed worldwide in 1973, while not comparable with the harvest of nearly two thousand million tons of cereals and root crops, are an important contribution to human diet and provide a livelihood for millions of fishermen. Protein is an essential part of what we eat, but is supplied only in small quantities by cereals and root crops, so that many people in developing countries suffer from a protein shortage.

Diets that include animal meat, pulses (peas, beans and similar vegetables) or fish can supply this deficiency, and the importance of the fisheries lies partly in that marine and fresh water together provide about 20% of the animal protein in the developing world. In the West, fish and shellfish provide an alternative source of high-grade animal protein, and smaller or less palatable fish are processed into meal and enter our diet after conversion to chicken, pork or beef. At present, almost all marine fish and shellfish are harvested by conventional fisheries, a method of exploitation which requires (but does not always receive) careful regulation to ensure an optimum yield. In future, new fisheries and fish farming may make an important contribution to the protein shortage in developing countries and, in developed countries, may provide an additional supply of high-grade fish.

This chapter is about marine production and the way in which we make use of it. We exploit the production of the land and sea in very different ways. As Table 1.1 shows, much of the harvest of the land comes directly from plants, and almost all is the result of husbandry of some kind.

Although the sea covers two-thirds of the globe, the minute plants that drift in the illuminated region close beneath its surface convert in total only half as much sunlight into the chemical energy of living matter as do the larger plants of the land. The reason for this is that the plants of the sea are small and relatively few, and this in turn means that we cannot crop the marine harvest in the same way as we gather in that of the land. Instead we take the fish which eat smaller animals which eat the plants and, as a result of the losses involved in this, the sea provides only a small part of human food. Appreciation of the potential for, and the limitations on, marine food production requires an understanding of the principles of marine production ecology.

Table 1.1 World harvests of certain foods in 1972.

	Million tons
Cereals	1368
Root crops	581
Pulses	44
Meat	110
Marine fish, crustaceans and molluscs	52
Fresh-water and migratory fish	12

Very little of the yield of the sea is farmed in any way and, indeed, as fishermen or collectors of shellfish or seaweed, humankind has advanced from the hunters and gatherers of the Stone Age only in the technical development of the tools used, and not in the management of the ecology of the marine species exploited. Thus it is not misanthropic to see our species simply as one of many predators of fish, and hence, as is done here, to apply to the fisheries certain general ecological ideas about predators and their prey.

The natural history of the sea

Life in the sea

Warm water is lighter than cold, and thus in regions where the sea surface absorbs much heat from the air or the sun, a layer of warm low-density water is found floating on colder denser water. In addition, light is relatively rapidly absorbed, even by the clearest seawater, and so the illuminated zone of the sea is comparatively shallow. Only here can plants

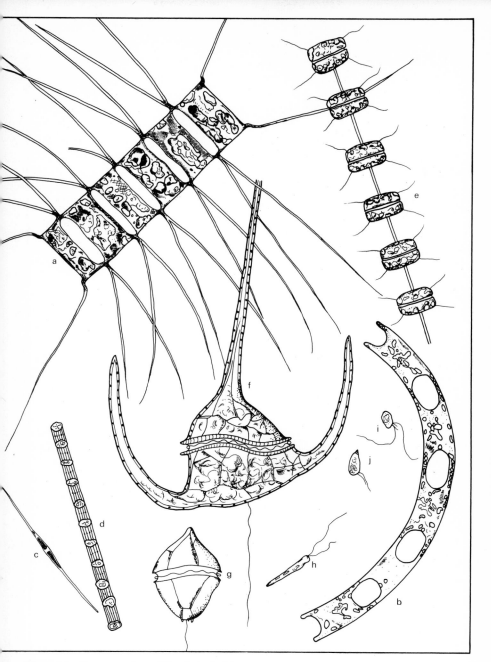

Figure 1.1 Some members of the phytoplankton.
All the phytoplankters shown here are common in British coastal waters, and all are drawn to the same scale. The long side of the illustration represents about 0·4 mm.
　　Diatoms: a, *Chaetoceros decipiens*; b, *Eucampia zodiacus*; c, *Nitzschia closterium*; d, *Skeletonema costatum*; e, *Thalassiosira rotula*.
　　Dinoflagellates (note the encircling and the trailing flagella): f, *Ceratium tripos*; g, *Gonyaulax spinifera*.
　　Flagellates: h, the euglenoid *Eutreptiella hirudoidea*; i, j, two small unidentified forms— j is probably a member of the golden-brown algae.

grow, and thus many of the processes to be described in this chapter are restricted to this superficial zone of warm illuminated water.

Most of the sea is deep, dark and cold. Here there are no plants, and the scattered animal life depends on food sinking from above. Although, for example, the average depth of the Atlantic Ocean is about 3300 metres, the beds of this and other oceans are not flat; they contain both deep trenches, and peaks and ridges. Some of these marine mountains rise above the sea surface, e.g. the Azores in the Atlantic or the Hawaiian archipelago in the Pacific. Around the land masses are relatively shallow areas, the *continental shelves*, which have depths less than 200 metres.

The waters of the sea are always in motion. Surface currents like the Gulf Stream transport large quantities of warm or cold water away from their original latitudes at speeds of several kilometres a day. There are also slower currents in the deep sea, their movements channelled by the shape of the ocean basins. Local but large increases in fertility are caused where deep waters rise to the surface in regions of *upwelling*. As we shall see later, these regions are the most productive in the sea. Next in order of fertility is the water over the continental shelves, where the closeness of seabed to surface aids the return of essential nutrients to illuminated waters.

Life in the sea takes a variety of forms. Much of it is small and drifts passively, apparently at the mercy of wind, tide and current. This is the *plankton*. Some of the animals of the *zooplankton* are illustrated in figure 1.2, and some of the minute simple plants of the *phytoplankton* on which they feed are shown in figure 1.1. As already mentioned, the phytoplankton are confined to the surface waters of the sea. Many of the zooplankton, however, exhibit the behaviour known as *vertical migration*. During the day they are found below the illuminated zone. At dusk they swim upwards to feed on the phytoplankton of the surface waters, returning to the deeper water at dawn. Such movements, of only a few hundred metres, do not contradict the idea of the zooplankton as drifters, for such small animals can make no significant horizontal progress. In this they differ from larger animals like fish that can swim actively from place to place. These active animals and drifting plankton, all of which live surrounded by water, are called *pelagic* to distinguish them from the *benthic* organisms that live on or in the sea floor. The latter category includes shellfish, and many kinds of worms and smaller animals.

There are also the large benthic plants, the seaweeds and the eelgrasses, which are confined to shallow waters by their need for illumination. Finally there are the marine *decomposers*: bacteria and other microorganisms which play a most important role in the ecology of the sea, and which live in the bottom muds or attached to waterborne particles.

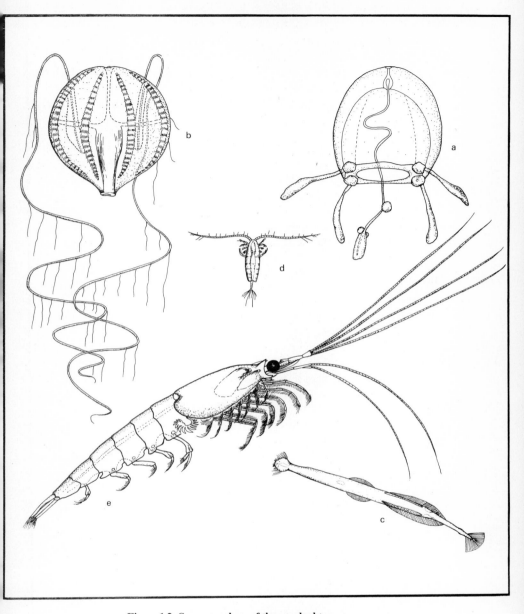

Figure 1.2: Some members of the zooplankton.
All the animals shown here are common in British coastal waters and are drawn to the same scale. The long side of the illustration represents about 5 cm.
a, the hydromedusan jellyfish *Sarsia* sp; b, the comb-jelly *Pleurobrachia pileus*; c, the arrow-worm *Sagitta elegans*; d, the copepod *Calanus finmarchicus*; e, the euphausiid *Meganyctiphanes norvegica*.

Food chains and nutrient cycling

A *food chain* describes a sequence of organisms that feed one on another, the ultimate source of their sustenance being always the organic matter formed from simple inorganic compounds—in particular, carbon dioxide—by plants exposed to sunlight. *Photosynthesis*, the sequence of reactions in which organic compounds are formed, is a process involving the capture and conversion of solar energy, which becomes temporarily locked up as the chemical potential energy of these organic compounds. Their subsequent oxidation to carbon dioxide in *respiration* liberates this energy for the use of the plants or the animals that have eaten them. All reactions taking place in living organisms are governed by normal chemical and thermodynamic laws. In particular, all reactions involve the loss of some energy as heat, and this cannot be reused. Thus, when stated in terms of a transfer of chemical potential energy, food chains are necessarily unidirectional. There can be no biological perpetual motion. Considered in terms of material flow, however, each chain is part of a cycle. Carbon in organic form is returned through the respiration of plants, animals and micro-organisms to carbon dioxide in the atmosphere, and to the various forms of dissolved inorganic carbon in the oceans. In the sea there is rarely a shortage of inorganic carbon for photosynthesis, and consideration of the non-living part of the carbon cycle is thus not of vital importance for marine ecology. Conversely, nitrogen and phosphorus are important examples of elements which are necessary components of protoplasm, and are often in short supply in the sea.

In the simplest marine food chains, phytoplankters are eaten by herbivorous zooplankters (e.g. copepods), and these in turn are eaten by zooplanktonic carnivores (e.g. arrow-worms) or by fish. The complexity of most feeding relationships is only partly instanced by the simplified food chain of the herring shown in figure 1.3. Two consequences of this complexity are worth noting. First, a varied diet allows flexibility; if food of one sort becomes scarce, it is possible for an animal to eat more of another kind. Secondly, the plexus of feeding relationships amongst marine organisms implies that natural or artificial changes act not only on isolated parts of the marine ecosystem, but may affect links in the food chains of many animals.

The food web shown in the figure is purely pelagic, except in so far as some of the larvae come from and will eventually rejoin the benthos. At least three *trophic levels* can be distinguished: that of the *primary producers*, the plants that utilize the energy of sunlight to convert inorganic into organic compounds; that of the *herbivores*, the animals that feed directly on the phytoplankton; and that of the *carnivores* that feed

MARINE PRODUCTION

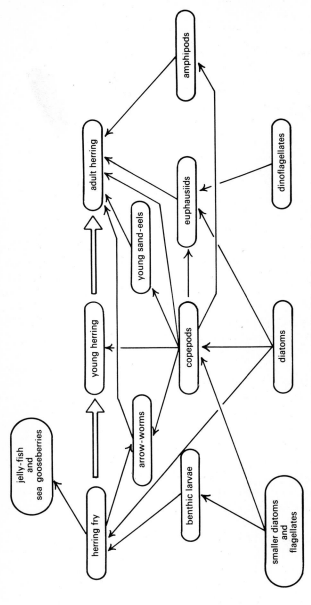

Figure 1.3: Simplified food chain of the herring.
"Fry" refers to larval and youngest post-larval herring. The amphipods mentioned in this diagram are members of a group of planktonic and benthic crustaceans that include the sandhoppers sometimes found in rotting seaweed.

on the herbivores. When feeding on copepods, adult herring act as *primary carnivores* on the third trophic level, two steps away from the primary producers. A fish that eats herring would be a *secondary carnivore* on the fourth trophic level. As figure 1.3 shows, animals sometimes feed on more than one trophic level. Thus, when eating sand eels, the herring is also acting as a secondary carnivore.

At each step in the chain some food sinks out of the surface layer of the sea. Dead and partially eaten phytoplankton and zooplankton, faecal pellets of zooplankton and fish, and the moulted exoskeletons of crustaceans provide a source of food for many animals. The deeper down in the sea, the more important is this food supply. In addition, the daily vertical migration of many zooplankters and fish transfers live food downwards. Live or dead, what is not eaten or metabolized on the way provides the food source of the deep-sea benthos. In coastal waters, this supply is augmented by terrestrial organic detritus brought to the sea in rivers, and by production by benthic plants.

Pelagic and benthic bacteria are very important in nutrient cycling. They convert the nitrogen and phosphorus locked up in organic compounds to soluble nitrates and phosphates and, as most of the process takes place outside the bacterial cells, much of the inorganic nutrient thus produced is returned directly to the water and becomes potentially available for uptake by phytoplankton. Although some nutrients are recycled more rapidly as a result of the metabolism and excretion of zooplankton and fish, it is nevertheless bacteria that supply most of the nutrient regenerated in the sea. Because of the downward movement of organic material, most regeneration takes place on the seabed, or in the deeper water beneath the illuminated surface layer.

In the benthos, the rain of detritus that is not directly eaten by large shellfish and other animals is initially metabolized by bacteria and Protozoa. Benthic food chains then proceed by way of smaller animals to the larger worms and other carnivores. At the top of the food chain stand *demersal* (bottom-living) fish, such as plaice and cod, that feed mainly on larger crustaceans, worms and molluscs. In some areas, plaice have a habit of biting off the protruding siphons of buried bivalves, leaving their owners otherwise unharmed and apparently able to regenerate their missing organs!

The highest level of many marine food chains is occupied by seabirds and man. These, together with the fish that migrate from the sea to fresh water at some stage in their lives, serve to transfer food from sea to land. Only a small part of marine production takes this route; some of the associated nutrients are returned directly in seabird droppings.

Marine production ecology

Primary production
The term *primary production*, referring to the rate of formation of organic matter by the plants of the plankton, must be distinguished from *standing crop*, which refers to the amount of phytoplankton present at any one time. The example of a regularly mown lawn may make this distinction clearer. Mown once a week throughout the summer, the lawn remains properly short, and the standing crop, the amount of grass on the lawn, remains more or less constant. The cuttings emptied from the mower represent the weekly production of the lawn, and may, in the course of a summer, add up to much more that the weight of grass on the lawn at any one time. The case of the phytoplankton is analogous: zooplankton grazing is equivalent to mowing, and in many instances keeps the standing crop of phytoplankton down to a low and roughly constant level. The primary production, however, can sum in a year to a total that is many times the average standing crop.

In dealing with quantitative matters the precise definitions of, and the units used for, primary production and standing crop become of great importance. It is clear enough that the standing crop of grass on our lawn can be measured by digging it all up, weighing, and dividing by the area of the lawn. The weight of fresh grass, however, includes all sorts of variable components, and it is thus better to express the grass standing crop in terms of its organic carbon content. Carbon forms a relatively constant proportion, about 40%, of organic matter, and is the best measure of food or potential energy content of the grass. Very roughly, it is about 10% of live (wet) weight of most plants and animals.

For the purpose of comparison with the standing crops of land plants, that of phytoplankton is also best expressed in *grams of carbon per square metre*. Whereas the former plants are attached to the surface to which their measurement is referred, the plants of the plankton are dispersed through a water column that may be several hundred metres deep. The standing crop of phytoplankton thus refers to all the plant carbon in a water column of area one metre square, extending from the surface to either the sea-bottom or to the lower limit of penetration of light bright enough for photosynthesis.

The rate of primary production is the rate at which new organic matter is added to the existing plant standing crop. (No account is taken of possible losses due to grazing and other factors external to the plants.) It is expressed in terms of *grams of carbon per square metre*, either *per day* or *per year*, and thus, in the case of phytoplankton, refers to the total production in a one-metre-square column of water. Primary production may be considered as the product of standing crop and *photosynthetic rate*, thus:

Primary production rate (g carbon m^{-2} day^{-1})
= standing crop (g carbon m^{-2}) × photosynthetic rate (day^{-1})...(1)

The units of photosynthetic rate *per day* indicate a relative rate, i.e. one expressed in terms of grams of new organic carbon formed by photosynthesis per gram of existing plant carbon per day. In this way it is possible to compare the photosynthetic rates of plants of different sizes.

Photosynthesis

Almost all plants contain the green pigment *chlorophyll a* which is central to photosynthesis. When chlorophyll absorbs light of the right colour, the result is the breakdown of water molecules into oxygen and hydrogen, the latter reducing carbon dioxide to form simple carbohydrates. These serve both as an energy source and, if the nutrient elements already mentioned are available within the cell, as the basis for the synthesis of the more complex compounds that make up protoplasm. The uptake of inorganic carbon and its photosynthetic conversion to organic form is often called *fixation*.

The rate at which new organic matter is produced as an immediate result of the absorption of light energy is the *gross* rate of photosynthesis. At the same time as it is fixed in photosynthesis, however, carbon is lost by respiration, and the *net* rate of photosynthesis takes into account these respiratory losses.

Gross photosynthetic rate depends on the chlorophyll content of plants, and phytoplankton differ from land plants in having a relatively high

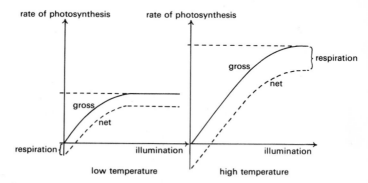

Figure 1.4 Photosynthesis and illumination at two temperatures.
Whereas photosynthesis under dim illumination is directly proportional to the amount of light absorbed, the maximum (light-saturated) rate of photosynthesis is dependent on temperature. Because of the increase in respiration, the difference between gross and net photosynthesis is greater at the higher temperature.

chlorophyll content. All other things being equal (which they rarely are) micro-algae can thus photosynthesize at a relatively faster rate.

The rate of photosynthesis is also controlled by light and temperature, as shown by figure 1.4. Gross photosynthesis at low illuminations is unaffected by temperature, whereas the *light-saturated rate* of gross photosynthesis under high illuminations increases with increasing temperature. So also does respiration.

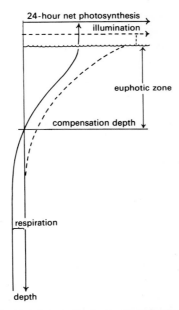

Figure 1.5   Net photosynthesis at various depths in the sea.
The compensation depth (at which 24-hour respiration equals daytime gross photosynthesis) marks the bottom of the euphotic zone.

Light is absorbed by seawater and by the particulate material in the sea. At some depth, therefore, the total gross photosynthesis carried out by phytoplankton during the hours of daylight is just balanced by the respiration that occurs during both day and night. This is the *compensation depth*, at which 24-hour net photosynthesis is zero. Below this depth phytoplankton cannot long survive; above it is the *euphotic zone*, the only part of the sea in which plant growth is possible, and which varies in thickness from several hundred metres in the clearest ocean waters to a few metres in turbid coastal waters.

Figure 1.5 illustrates these concepts in their simplest forms. One amongst many complications should be mentioned. Because the upper layers of the sea are often in vertical motion, a phytoplankter is unlikely to remain long at one depth. Instead, it may be carried down below the compensation depth and returned nearer the surface. It is thus the average rates of gross photosynthesis and respiration over 24 hours that are important.

Standing crop

Table 1.2 shows that, with the exception of seaweed and eel-grass communities, which are comparable with terrestrial grasslands, the oceans support an average weight of plants that is very much less than on land.

Table 1.2 Typical standing crops of various terrestrial and marine plant communities.

Type	g carbon m^{-2}	g chlorophyll m^{-2}	carbon/chlorophyll ratio
Temperate grassland	600	1	600
Temperate deciduous forest	12 000	2	6000
Densest tropical forest	80 000	12	6700
Temperate intertidal seaweed	400	4	100
Subtropical eel-grass bed	600	3	200
Coral reef algae	250	1	250
Phytoplankton of temperate coastal waters; euphotic zone:			
Winter minimum	0·1	0·002	50
Spring maximum	10	0·3	30
Summer mean	2	0·04	50
Phytoplankton of tropical waters, euphotic zone:			
Sargasso Sea, minimum	0·05	0·0005	100
Equatorial Pacific, mean	1	0·01	100

This is somewhat surprising at first sight, particularly as, owing to their relatively greater chlorophyll content, phytoplankters can grow considerably faster than can land plants. It might therefore be supposed that the most important restrictions on standing crop are those that directly control growth. In circumstances where the size of the standing crop is rapidly changing, growth-controlling factors are indeed important. In many cases, however, a rough equilibrium prevails between the rates of standing crop gains from net photosynthesis and losses due to grazing; and apart from winter in high latitudes, a shortage of nutrients may be considered the most important factor limiting standing crop.

Recent work with phytoplankters grown under controlled laboratory conditions has shown that these algae tend to a constant chemical composition. Such constancy is not always attained, and indeed our knowledge of regulatory mechanisms derives from experiments in which considerable variations are caused in composition. Nevertheless most healthy phytoplankters seem to have a make-up in which the ratios of atoms of carbon, nitrogen and phosphorus are roughly 100:15:1.

This approximate constancy is the result of the dependency of growth on cell nutrient concentration. Should nitrogen:carbon or phosphorus: carbon ratios decline, due to the photosynthetic fixation of additional carbon, growth rate slows.

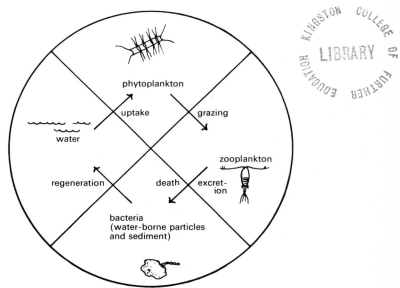

Figure 1.6 The nutrient cycle in the sea.
Very much simplified. Under equilibrium conditions, nutrient passes at a constant rate from one quadrant to the next.

Although phytoplankton take up nutrients from the sea at a rate that depends on outside concentration rather than on internal shortage—and indeed if presented with high concentrations of external nutrients, can take up much more than is immediately utilized in growth—there is nevertheless a tendency towards an equilibrium in which nutrients are taken up at about the same rate as they are used in growth. Figure 1.6 shows the uptake/growth equilibrium as part of a wider nutrient cycle that also tends to approach an equilibrium. Such an equilibrium, in which nutrients pass through the quadrants at approximately constant rates, is most nearly attained in the surface waters of tropical oceans where small amounts of nutrients are rapidly recycled as a result of grazing and bacterial regeneration, and where there are no significant seasonal changes in productivity. Conversely, in temperate latitudes, low phytoplankton standing crop during winter causes a bottleneck in nutrient uptake across the water/phytoplankton boundary, so that dissolved nutrients increase in concentration in the euphotic zone.

Seasonal and regional variations in primary production
The factors that control productivity are to some extent inter-related. Thus the euphotic zone is often shallow in nutrient-rich areas, because plankton and detritus decrease the transparency of the water. Conversely, in infertile waters a deep euphotic zone makes up to some extent for a low standing crop. Productivity thus varies less than might be expected between different parts of the oceans. It is possible, however, to contrast waters less than about 30° from the equator, where production is relatively constant throughout the year, with the seas of higher latitudes, where there are often pronounced seasonal cycles. It is also possible to contrast the shallow water over continental shelves with the deep waters of the high seas. Except in localized regions of upwelling, the latter are relatively unproductive.

Production in the equatorial oceans is both low and relatively constant. This is the result of the existence throughout the year of a *thermocline*, a region of pronounced temperature change at a depth, in typical cases, of several hundred metres. Above the thermocline is a layer of warm—and therefore less dense—water; below is cooler denser water. Because of the density difference the two layers do not mix, and the thermocline thus forms almost as good a floor to the surface water layer as does the sea bottom in shallow waters. An important difference is that the nutrient regeneration that occurs on the sea bottom takes place to a much lesser extent at the thermocline. Because dead phytoplankton sink, and because of downwards vertical migration by zooplankton after they have fed, nutrients are transported downwards and out of the euphotic zone.

Where the thermocline occurs below the compensation depth, the euphotic zone is considerably depleted of nutrients, although, as mentioned previously, remaining nutrients are rapidly recycled owing to high temperatures and illuminations. Phytoplankton concentration is low but, when expressed in relation to chlorophyll, the rate of gross photosynthesis is high, owing to high temperatures and illuminations. The euphotic zone is deep, owing to the clarity of the water, and the standing crop summed over the water column might be 1 or 2 grams of carbon per square metre. Multiplied by an average *net* photosynthetic rate of 0·1 per day (low due to relatively low algal chlorophyll content and the high rate of respiration resulting from the warmth of the water), daily net production averages about 100 milligrams of carbon per square metre.

Seasonal variability is unimportant, and annual primary production is probably between 10 and 50 grams of carbon per square metre. Relatively little work has been carried out on production under these conditions, however, in spite of their occurrence over a large part of the oceans, and these production estimates must therefore be treated with caution.

Better known is the production ecology of the temperate seas of the continental shelves around Europe, North America and Japan. The existence of an annual cycle of production in these waters has long been recognized. Dull and stormy winter days result in a shallow euphotic zone and a deep layer of mixed water. The consequence of mixing is that many phytoplankters spend much of their time below the compensation depth. Mean net photosynthetic rate is thus very low and can be negative if, on average, respiration in the mixed water layer exceeds gross photosynthesis. Winter water samples hence show low phytoplankton standing crops.

Meanwhile, nutrient regeneration continues in deeper waters and on the bottom, and mixing brings the regenerated nutrients back to the surface. As spring approaches, increasing light and decreasing storminess extend the euphotic and diminish the mixed zone. The mean rate of photosynthesis increases and there follows a period of rapid phytoplankton growth, culminating in the greatest standing crops of the year. This annual occurrence is the *spring phytoplankton increase*, and the large crops are possible because of the large amounts of nutrient available at this time.

By the middle of spring almost all of the dissolved nutrients in the euphotic zone have been taken up by the phytoplankton. Various factors bring about the decline of the standing crop: these include grazing by zooplankton and the sinking of algae suffering extreme nutrient depletion.

Whatever the cause of the decline, the tendency thereafter is towards a rough equilibrium that persists during most of the summer. Much nutrient is unavailable inside zooplankton and in detritus that has sunk beneath the euphotic zone. Often the development of a summer thermocline completes the isolation of surface from deep water. Regeneration continues below the thermocline, and the first storms of autumn may, by mixing, bring enough fresh nutrients into the euphotic zone to allow the phytoplankton a short-lived autumnal increase. This soon ends, and with shortening days the decrease in the average photosynthetic rate results in a decline of phytoplankton abundance to the level of the previous winter.

During the productive season the phytoplankton are thus nutrient-limited, whereas during winter they are limited by insufficient light. Annual primary production is about 100 grams of carbon per square metre; more details are given in figure 1.7 and Table 1.3.

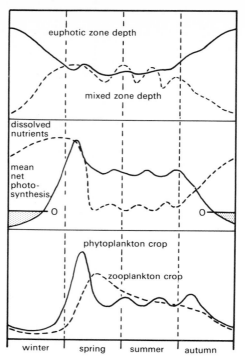

Figure 1.7 The annual cycle of production and standing crop in the waters over a continental shelf in temperate latitudes.

On a world scale, the regions of the oceans may be grouped into four types on the basis of their productivity. As has already been discussed, the equatorial oceans (between 30°N and S) generally have low productivity as a result of the euphotic-zone nutrient depletion that results from marked thermal stratification. Temperate and sub-polar oceans have somewhat higher, but seasonally variable, productivity, being limited by light in winter and nutrients in summer. Seas of continental margins have a high productivity (greater than 100 grams of carbon per square metre per year) for a number of reasons, including the proximity of the bottom to the euphotic zone rendering nutrient regeneration more rapid, the enrichment of nutrients from terrestrial sources, and the contribution of benthic plants. Finally, some small regions of the open ocean have a very high productivity due to the upwelling of nutrient-rich deep water, e.g. parts of regions of the Iceland-Faroes Ridge, the coastal waters of Peru, and the Antarctic Convergence show productivities of more than I gram of carbon per square metre per day on many days of the

MARINE PRODUCTION

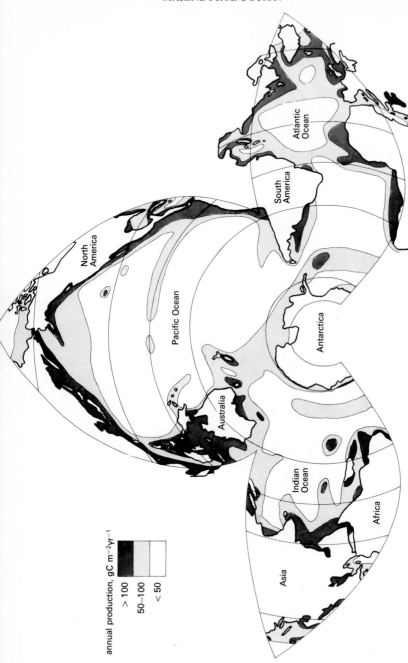

Figure 1.8 Global distribution of primary production.
Regions of highest productivity are located over continental shelves or where deep water comes to the surface, as along the Pacific coast of South America or at the boundary between Antarctic and temperate water.

year, and may have annual productions exceeding 200 grams of carbon per square metre. It must, however, be emphasized that areas of such sustained high productivity are very limited in extent.

Global production

The most recent estimates of the primary production of the oceans give a world total of about 2×10^{16} grams of carbon per year, or a mean rate of about 60 grams of carbon per square metre per year. This represents an efficiency conversion of solar to fixed chemical energy of only 0·03%. The land, with half the area, supports a primary production twice that of the sea and in excess of 4×10^{16} grams of carbon per year. This represents a mean areal rate of nearly 300 grams of carbon per square metre per year and a conversion efficiency of 0·13%. These figures suggest two questions. Why are mean conversion efficiencies so low? And why is the average rate of marine primary production only about a fifth of that on land?

A *photosynthetic conversion efficiency* of 0·03% means that of the solar energy reaching the upper atmosphere over the sea, only three parts in

Table 1.3 Typical net rates of primary production of various marine and terrestrial plant communities.

Community	mg carbon m^{-2} day^{-1}	g carbon m^{-2} year^{-1}
Phytoplankton		
Equatorial Pacific	10–200	35
North Atlantic	150 (summer)	75
	1000 (spring max.)	
North Sea	10 (winter min.)	100
	1500 (spring max.)	
	200 (summer)	
Atlantic in upwelling region off S.W. Africa	500–4000 (December)	300
Benthic Plants		
Subtropical eel-grass		1000
Temperate intertidal seaweeds	2000 (growing season)	350
Terrestrial Plants		
Desert, semidesert, tundra and alpine		30
Temperate grassland and forests		360
Tropical grassland and forests		580
Wetlands		800
Average terrestrial		300
Average marine		60
Average global		130

ten thousand are on average transformed by photosynthesis into the chemical potential energy of organic compounds. This must be contrasted with the theoretical efficiency of 28% for gross photosynthesis that is obtained in laboratory experiments carried out under the most favourable conditions. However, such high efficiency occurs only at low levels of illumination, and only if all the light is of the appropriate colour and is absorbed totally by photosynthetic and accessory pigments. Infra-red radiation, comprising about half the energy of solar radiation, cannot be used by plants in photosynthesis. Neither can ultra-violet or some visible light. Thus only about 40% of solar radiation is suitable for photosynthesis and, although this light penetrates seawater better than does the other 60%, some has already been absorbed by the atmosphere or reflected by clouds, and much more is absorbed by the water, by dissolved material and detritus, by zooplankton and by nonphotosynthetic parts of the phytoplankton. At high light intensities, photosynthesis takes place with reduced efficiencies, due to saturation.

Respiration accounts for part of the energy converted by gross photosynthesis, thus further reducing efficiency. The largest loss of efficiency, however, simply results from the low average standing crop of phytoplankton. Much of the available light is never absorbed by algae.

It is possible to compute the size of the phytoplankton standing crop that will most efficiently utilize available light, the upper limit to the theoretical phytoplankton concentration being that at which self-shading becomes severe. If global marine productivity were limited by light rather than by nutrients, the annual figure would probably be about ten times the present 2×10^{16} grams of carbon per year, i.e. a rate of about 600 grams of carbon per square metre per year and a conversion efficiency of 0·3%.

The growth of land plants is controlled less by shortages of nutrients and light than by shortages of water and the low winter temperatures of high latitudes. Greater nutrient availability on land leads to larger standing crops, and so to plant communities that absorb a greater proportion of incident light than do phytoplankton.

Although the greater efficiency of individual phytoplankters (the result of greater chlorophyll content and relatively smaller respiratory losses) goes some way to make up, it is not sufficient to compensate, for these great differences between planktonic and terrestrial standing crops. The most productive plant communities in the world are those of the attached marine plants of shallow waters and the terrestrial plants of many wetlands. Here neither nutrients, water nor, in low latitudes, light, limit production. Such areas are potentially of great importance for human exploitation of production, but unfortunately are both of limited extent and extremely susceptible to human interference.

Production at higher trophic levels

Primary production has been dealt with at some length because it is the photosynthesis of phytoplankton that provides almost all of the food for, and thus determines the production of, the higher levels of marine food chains. In general it is true that the regions of the oceans that have a high primary production also have a large zooplankton crop and an actual or potentially rich fishery.

The richness of the benthos depends partly on primary production and partly on the depth of the sea, and the scantiness of the deep-water benthos is the result of the consumption of much of its potential food before this reaches the bottom. The richest benthos occurs in shallow-water areas, either of high marine productivity or, as in estuaries, where there is an additional source of food provided by terrestrial detritus. Such areas of high benthic productivity are important in the life of demersal fish, as will be seen later in the case of the plaice. In the rest of this section, however, only pelagic food chains are discussed.

As already mentioned, phytoplankton is eaten by herbivorous members of the zooplankton, of which the copepod *Calanus* is a good example. Although it can pick larger plant cells selectively from the water, smaller phytoplankters are filtered *en masse* by the bristles on the mouthparts. When food is abundant, many of the plants are only partially digested during their passage through the copepod's gut, and the faecal pellets produced at this time are an important food supply to deeper zooplankton and to the benthos.

A considerable part of the food that is digested is used to provide energy for the copepod, leaving the remainder for conversion into *Calanus* tissue or eggs. Only this latter part is therefore available to the carnivores that prey on *Calanus*, and the *growth efficiency* of an animal at the second trophic level can thus be defined as

$$\frac{\text{amount of herbivore tissue} + \text{eggs produced}}{\text{amount of phytoplankton eaten}}$$

Experiments with *Calanus* show that under favourable conditions more than 30% of digested food is converted into body tissue and eggs. Under these conditions, however, only about two-thirds of food eaten is digested. Conditions in the sea are not often so good, and natural growth efficiencies for *Calanus* probably range between 5% and 20%, with an average of 15%.

If *Calanus* is typical of zooplanktonic herbivores, and if during the course of a year all phytoplankton net production is eaten by zooplankton, then, for example, in temperate seas with an annual phytoplankton production of 100 grams of carbon per square metre, the annual production of zooplanktonic herbivores might be 15 grams of carbon per square metre.

Although the standing crop of phytoplankton in these seas only exceeds 2 grams of carbon per square metre during the spring increase and occasional summer blooms, that of the zooplankton might be several times this during much of the spring and summer. The explanation is that copepods live longer than diatoms. Whereas the latter may divide once a day during productive periods, and new cells are likely to be eaten almost as rapidly as they are produced, *Calanus* requires several months to grow from egg to adult.

As a result of this, the abundance of zooplankton in temperate waters fluctuates much less than that of phytoplankton, but there is nevertheless a distinct annual cycle. Over-wintering *Calanus* lay eggs whose spring hatching coincides with the phytoplankton increase, as is true of many zooplankters. At the same time the zooplankton of coastal waters is enriched by the larvae of benthic animals. Peak zooplankton crop is attained shortly after the phytoplankton peak, and thereafter decreases due to the settlement of benthic larvae and the results of predation.

In addition to the general relation between mean zooplankton production and mean phytoplankton production, variations from year to year in the timing and extent of the spring increase can have important effects on the success of the zooplankton in each year. In general, what happens at any particular level in a marine food chain depends both on the events at the level below, which determine food supply, and those at the level above, which determine the amount of predation. All levels of a food chain are interlinked in this way, and the importance of these links will become even more apparent in the next section when the factors affecting the maintenance of exploited fish populations are considered.

Thus the effects of a poor and late phytoplankton increase may be felt all the way up the food chain, resulting in many deaths from starvation amongst the fish larvae that hatch in spring. Conversely, better-than-average survival of young fish can result from a good spring increase that coincides in timing with the hatching of zooplankton eggs and the release of benthic larvae, and which thus ensures a good food supply for the fish.

Ecological efficiency and potential fish production
As has already been illustrated, the terms *standing crop* and *production rate*, previously defined only for phytoplankton, can also be applied to zooplankton, benthos or fish. In each case the *standing crop* is the average weight of living animals beneath a square metre of sea surface, and the

production rate is the amount of new animal tissue produced each year, net of respiration but disallowing losses due to predation.

Whereas in many waters the standing crop of phytoplankton is less than that of the zooplankton, the crop of fish about equal to that of the zooplankton, and that of the benthos somewhat more, production rate must always decrease substantially with increasing trophic level. Thus a phytoplankton production of 100 grams of carbon per square metre per year might support a zooplankton herbivore production of only 15 grams of carbon per square metre per year. Of the remaining primary production, perhaps 30 grams of carbon per square metre per year will go into faecal pellets (potential food for the benthos) and perhaps 55 grams of carbon per square metre per year will be respired by the zooplankton (figure 1.9). In cases like this, where all phytoplankton production is eaten by zooplankton, the *ecological efficiency* of the zooplanktonic herbivores, given by the ratio

$$\frac{\text{net production of zooplanktonic herbivores}}{\text{net phytoplankton production}}$$

is the same as the growth efficiency of the individual herbivores, i.e. 15%.

In considering the potential yield of the sea, it is the production rate with which we must be concerned, not the standing crop obtaining at any one time. The importance of harvesting from the production of a fish stock, rather than from its standing crop, will be emphasized again in the next section and, as the above discussion and that on food chains in general have shown, the production of fish must be supported by, and be less than, the production of the organisms on which they feed. Thus if the feeding habits of the herring place it completely at the third trophic level, the potential yield of herring in a region of known primary production can be calculated from:

primary production × ecological efficiency of zooplankton herbivores × ecological efficiency of herring.

Applying the *Calanus*-derived value of 15%, potential herring production is thus $(0.15)^2$ ($= 0.02$) of primary production.

In fact the herring sometimes feeds at the fourth trophic level, so its potential production will be less than calculated, perhaps 0.01 of primary production. This assumes that all pelagic food chains lead to herring, which is not true. Its ecological efficiency is therefore less than 15%, and the predicted yield must be reduced accordingly.

In order to calculate the global production of fish potentially exploitable by mankind, it is necessary to estimate the ecological efficiencies both of these fish and of the steps in the food chain leading to them.

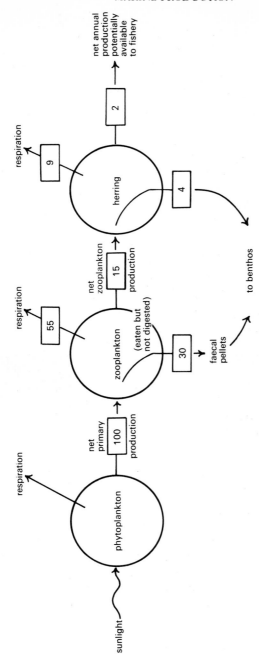

Figure 1.9 Food chain of the herring, to illustrate the idea of ecological efficiency.
The ecological efficiency of the zooplankton, for example, is the ratio

$$\frac{\text{net zooplankton production}}{\text{net primary production}} = \frac{15}{100} = 15\%$$

Work on terrestrial ecosystems suggests that ecological efficiencies of whole trophic levels probably average about 10%. On land, however, much plant production is not consumed by herbivores but enters the food chain indirectly and more slowly by way of decomposers. In the sea, plant production is usually consumed very rapidly, and the efficiency of this step in the food chain is thus the same as the mean growth efficiency of individual herbivores. Although all terrestrial production is eventually consumed, the incomplete utilization of primary production by herbivores renders the ecological efficiency of the plant/herbivore step in the food chain considerably less than the growth efficiencies of individual herbivores. Thus mean-trophic-level ecological efficiencies are probably higher in the sea.

On the other hand there is the problem of marine animals feeding at several trophic levels. Take the case of herring, feeding both at the third trophic level, on copepods, and at the fourth trophic level, on sand-eels, which also eat copepods. If feeding on copepods alone, both herring and sand-eels may show growth efficiencies of, say, 15% but when some of the copepod production reaches the herring indirectly through sand-eels, relatively more is lost in respiration. Thus herring growth efficiency, if still defined as

$$\frac{\text{herring production}}{\text{copepods eaten}}$$

will be lower. In general, the problems of animals feeding at more than one trophic level can be resolved by specifying the trophic level of the animal concerned as the lowest of the levels at which it feeds, and reducing its growth and ecological efficiencies accordingly.

These and other considerations suggest the use of an average of 10% for trophic-level ecological efficiencies in the sea. The fourth trophic level is probably typical of many of the fish in which we are interested, and production at this level is thus $(0 \cdot 1)^3 = 0 \cdot 001$ of primary production. Perhaps half of this production is in forms that are unlikely to be exploited, and thus, taking world primary production as 2×10^{16} grams of carbon a year, the global production of fish of potential importance to humans can be calculated as 1×10^{13} grams of carbon a year, or about 100 million tons wet weight.

A more realistic approach is shown in Table 1.4. It employs ecological efficiencies of 8–15% and takes account of differences between lengths of food chains and kinds of fish taken by fisheries in areas of different productivity. Highly productive areas of upwelling, such as that off the Peruvian coast, support dense shoals of planktonivorous fish. Although only a small part of the world's seas, these regions yield a large part

Table 1.4 Calculation of world total marine primary production and fisheries yield

Type of sea	Area (10^6 km^2)	Primary production (g carbon m^{-2} yr^{-1})	Total primary production (10^{12} g carbon yr^{-1})	Trophic level of exploited fish	Ecological efficiency %	Potential fish yield (10^6 tons yr^{-1})	Probable yield to conventional fisheries (10^6 tons yr^{-1})
Upwelling areas	0·6	200	120	2–3	15	70	50
Seas of continental margins	30·0	120	3600	3–4	10	110	55
Temperate and subpolar oceans	150·0	75	11300	4–5	10	36	15
Equatorial oceans	180·0	35	6300	5	8	3	1
All	361·0	60	22000	4	10	220	120

Note that, because of rounding errors, columns do not always add up to world total.

of the total fish catch. Much of the rest is landed from continental-shelf water of medium productivity, where both pelagic and demersal fish are exploited. These include both plankton feeders and fish eating other fish or benthic animals. In the infertile reaches that make up much of the high seas, the only fish worth exploiting are large secondary or tertiary carnivores, such as tuna.

Most other methods of calculation estimate the potential world harvest of exploitable fish to be between 50 and 200 million tons wet weight a year. The most likely value would seem to be about 100 million tons a year, as suggested by the two methods of computation given here.

The fisheries

Introduction

At the end of the Second World War, world landings of marine fish, molluscs and crustaceans were slightly less than 20 million tons. They rose steadily throughout the fifties and sixties, reaching almost 60 million tons in 1970. A number of fisheries made major contributions to this rise in landings. The traditional fisheries of the industrialized nations of the northern hemisphere expanded up to or beyond estimated sustainable yield. In addition, some of these nations began fishing very far from home. Thus the Japanese now fish with longlines for tuna on the high seas all over the world, and Russian trawlers now take a substantial quantity of Cape hake from the Southern Atlantic. The fishery for the Peruvian anchoveta increased from almost nothing to more than 12 million tons, and fisheries were developed in other upwelling regions. Finally, other developing nations have begun or expanded their own modern fisheries; a good example is the trawl fishery in the Gulf of Thailand.

In many of these cases there is a danger that over-fishing is seriously affecting the abundance of fish stocks, and in fact world landings have declined by about 6 million tons since 1970. Of major importance has been the drastic decline in the Peruvian anchoveta fishery (down to 2 million tons in 1973), but there have also been reductions in landings from other heavily exploited fish stocks, as in the case of the herring of the North-East Atlantic. Thus it appears that while we are at present exploiting by means of the fisheries a large part of the estimated realizable fish production of 100 million tons a year, we are as likely to reduce the catch as to increase it by any increase in the intensity of fishing. To understand this, we must first understand general principles and know something about the life history of the fish concerned.

The herring

The herring is one of the world's most important sea-foods, accounting for 5% of the weight of all marine fish landed in 1973. The North-East Atlantic provides more than half of these herring and, with a value of nearly £9 million in 1973, the fish made up 28% by weight of landings by Scottish boats. Formerly largely taken from the North Sea, increasing numbers now come from the North and West of Scotland, and it is the lives of the herring in these two areas that will now be described.

The herring is a pelagic fish and, as figure 1.3 shows, it feeds mainly on zooplankton. At night the herring follow their prey as the latter migrate towards the surface. It was thus that the fish were once taken, caught by their gills as they swam into the mesh of drift nets put out like long fences in the water. During their breeding season in particular, herring assemble in large shoals, and the drift net fishery was most intense at this time.

Nowadays herring are fished both by day and night, using many different sorts of nets: purse seines, which enclose a shoal and are than closed from below; Danish seines, in which the fish are herded into the net by the closing of a circle of rope on the seabed; bottom trawls; and fast pelagic trawls that are used particularly in deeper waters.

Unlike most of the major food fish, a female herring produces a relatively small number (between 10,000 and 80,000) of yolky eggs which she lays on the seabed. A number of herring spawning grounds, which are small and scattered, are known around the British Isles. The eggs are mostly laid in autumn. They stick to the bottom and hatch from one to four weeks later to produce larvae that float upwards to become members of the plankton—not feeding, but subsisting off the remains of their egg-yolk, which is contained in a bag beneath their stomachs. When this is used up, they begin to feed first on phytoplankton and then, as their own size increases, on small zooplankton.

At the same time they provide food for carnivorous members of the zooplankton: for jellyfish, sea gooseberries and arrow-worms, as is shown in figure 1.3. This is a crucial stage in the life of the herring. If there is a shortage of phytoplankton and small zooplankton, perhaps because the spring increase is delayed that year, the herrings' yolk supplies are exhausted before live food becomes available, and many larvae die of starvation. In addition, the presence of large numbers of sea gooseberries or arrow-worms may drastically reduce the ranks of the young herring. Thus the proportion of fish surviving to the end of their first summer varies greatly from year to year. As will be discussed later, this proportion is of great interest to fishery scientists, but it is dependent on so many factors that it is at present impossible to predict accurately.

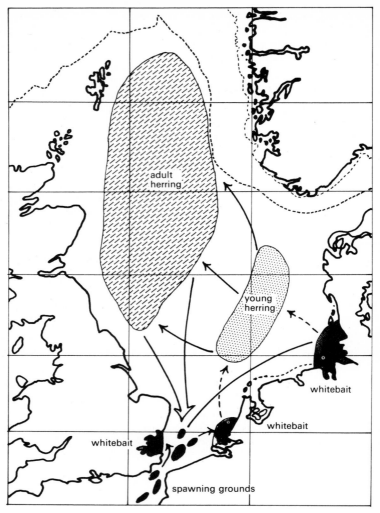

Figure 1.10 The movements of young and adult fish belonging to the Downs stock of North Sea herring. The 200-m depth contour is also shown.

One of the most interesting features of the herring's development is the movement from spawning to nursery to adult feeding areas. In the North Sea, water currents bring the young herring to inshore regions and estuaries by the time that they are six months old, and here they stay for up to a year before moving offshore. The fish become mature at two to three years of age, and join the adult population in the northern

North Sea. These movements are shown in figure 1.10 for fish spawned in the Eastern Channel and Southern Bight. These fish, which return to the same area to breed, represent a *stock*. Together with two other stocks which spawn further north, they make up the North Sea herring population.

In the case of the Hebridean herring, the nursery grounds are the firths and lochs of the west coast of Scotland, and probably the eastern coastal waters of the Moray Firth. Adult herrings are found in the Minch, in the waters to the north and west of the Hebrides and to the north-west of Ireland, and in the Firth of Clyde.

Although man is presently the most important predator of the herring of the North-East Atlantic, many other animals feed upon it. The smaller toothed whales and porpoises, sea-gulls and gannets, all take herring, and many more are eaten near the bottom by cod and other demersal fish.

The adult herring thus stands part way in a food chain that proceeds from phytoplankton via copepods, euphausiids, amphipods and sand eels, to the herring, and to its predators.

In the mid-sixties herring landings from the North Sea and Skagerrak averaged about 1·2 million tons a year. Calculating backwards from the third trophic level with a growth efficiency of 10%, this represents a primary production of $1\cdot2 \times 10^{13}$ grams of carbon per year. Average production in the North Sea is about 100 grams of carbon per square metre per year, and the area that supports the herring fishery is about 400,000 square kilometres. Total primary production is thus about 4×10^{13} grams of carbon per year, of which the herring landings account roughly for three-tenths. The importance of the herring in the marine food chain is thus clearly demonstrated, and the effects of the fishery shown likely to be extremely significant.

The plaice

South of a line joining Buchan Ness in Aberdeenshire to the northern tip of Denmark, the North Sea is less than 100 metres deep, and contains shallows, e.g. the Dogger Bank, that come to within 30 metres of the surface. This is the most productive part of the North Sea and supports the richest benthos. From it in 1973 were taken 120,000 tons of plaice, the major part of the world landings of this fish. In Scotland in 1973 the plaice accounted for some 5000 tons, or 1% of the total landings. However, its market value is considerably higher than that of the herring, and these fish fetched about £1 million.

As the distribution of their fishery implies, plaice are shallow-water demersal fish, feeding on benthic worms and bivalve molluscs. Since many of the worms are carnivorous, and many of the bivalves feed near the top of the detritus-based food chains, the plaice must be set at the fourth trophic level. In addition, they are in competition for their food with other species of demersal flat and round fish, and with carnivorous invertebrates, and thus it is not surprising that the plaice fishery yields only about a tenth of the catch of the herring fishery in the same area.

Plaice spawn in winter and, unlike the herring, but like most commercially important fish, the female lays a large number (about half a million) of floating eggs. The greatest concentrations of eggs are found in the centre of the Southern Bight, i.e. between the mouth of the Thames and the coast of Holland. They hatch into larvae that have the long thin shape typical of the young of many fish, and which subsist, like the larvae of the herring, at first on the remains of their egg yolk, and then on small members of the plankton.

Meanwhile water currents carry them towards the Dutch coast, and here, about a month after hatching, they metamorphose into miniature flatfish, one to two centimetres long, and settle to the bottom to feed on the benthic crustaceans of these coastal "nursery" areas. The survival of the larvae up to metamorphosis depends on many factors, including the amounts of planktonic food available and the abundance of predators. In addition, the strength of the eastwards flow in this part of the North Sea is crucial. If too slight, the larvae will metamorphose before they reach the best nursery grounds; if the flow is too strong, the change of life will occur too late. Thus the success of reproduction varies greatly from year to year, as in the case of the herring.

The young plaice spend a year or more on the coastal nursery grounds of the Wadden Sea, between the Frisian Islands and the Dutch, German and Danish coasts, and then gradually move out towards the middle of the North Sea to join the adult stock when they are between 3 and 5 years of age. Here they are fished with otter and beam trawls and Danish seines.

Less abundant stocks occur to the west of Scotland, where young plaice are found close inshore in many sandy-bottomed bays and sea lochs. As with the North Sea plaice, fish move offshore as they grow older.

The coastal regions that provide nursery grounds for the young of plaice, herring and other fish are unfortunately also those parts of the sea most influenced by human activities. Pollution provides one example of such human influence, and the effects of enclosure another. Thus the success of the Dutch in creating polders in parts of the Wadden Sea might not have been entirely beneficial, for it has reduced the feeding area available to young plaice. Changes in the environment in one part of the North Sea might thus influence the fishery in another part and, although there is to date no evidence that the Dutch enclosures have affected the plaice fishery, it is vital to determine the importance of various parts of the Wadden Sea as plaice nurseries.

Prudent predators and efficient prey
The idea of a prudent predator is of an animal that is in balance with its prey, taking no more than can be replaced by the reproduction of the prey and the growth of its young. Except when the predator is a human, the terms *prudent* and *efficient* imply no volition, but describe situations resulting from the natural selection of predators whose instinctive behaviour causes them to feed thus advantageously. Indeed, in many cases the appearance of prudence is as much as a result of the complexity of the food web as of the behaviour of individual predators. In cases where laboratory experiments have been performed using one species of predator and one of prey, the result is almost always a series of reciprocal ups and downs in the numbers of each, ending in the extinction of the predator and sometimes also of the prey.

The complex food webs that exist in many natural ecosystems, however, tend to damp out such oscillations. This is because most predators can find another prey if their normal food becomes short, and because most organisms suffer from a variety of predators, and therefore do not experience great relief at the disappearance of any one of them. Natural ecosystems thus tend towards equilibrium in the relations between trophic levels. Indeed, production and consumption must always in the long run balance, for animals cannot eat more than plants produce, nor can plants grow without animals and micro-organisms to recycle nutrients.

Equilibrium is only a tendency, however, and one that applies less strongly to individual species. History and the scientific literature contain numerous examples of short- and long-term natural changes in the abundance of many commercially important fish. For example, the downfall of the Hanseatic League might have been partially due to the decline in the stocks of the Baltic herring which supported the medieval economy of these towns. Human fishing effort, if prudent, should therefore be directed so as to supplement the processes stabilizing fish populations, and this is yet another reason for wanting to know as much as possible about the workings of marine food chains and about the physical and biological factors responsible for natural variation in primary and secondary productivity.

The concept of *prey efficiency* can best be illustrated in relation to a fishery that is at equilibrium. Indeed, only a prudent predator can have a prey that is efficient, i.e. one that is exploited so as to give the greatest sustainable yield and, in the most general sense, the most efficient transformation of primary production into food available to the predator. The question is, however, under what equilibrium conditions does a fishery produce the greatest sustainable yield?

Figure 1.11 demonstrates that this occurs when there is a moderate amount of fishing. When there is little fishing, fish are abundant and a small amount of effort produces a large catch. When there is much fishing, more work is required to catch the same amount of fish, and the *catch per unit effort* can thus, if properly estimated, be considered an index of the fish standing crop. When the crop is high and fishing effort low, only a small proportion of fish are taken by the fishery; the rest go to natural predators. Because this is an equilibrium in which consumption and production are equal, annual *fishing* plus *natural mortality* must equal annual net production by the fish.

At a higher intensity of fishing a smaller standing crop results in a reduced annual net production of fish, but most of this production goes to the fishery. The greatest yield to the fishery occurs at some intermediate fishing effort, when the standing crop is not overly reduced, but where a good proportion of production is nevertheless taken by the fishery.

Figure 1.11 shows that a certain amount of fishing is actually good for the fish stock, causing an increase in net production. The reasons for this are complex. One is that at high standing crops the fish require relatively more of what they consume in order to provide energy to seek for food that is relatively scarce as a result of demand by many fish. Another reason is that under-fished populations contain many old fish, which grow less efficiently than young fish. These effects produce the result shown in the part of the diagram labelled *production: standing crop*, and have been observed, for example, in the North Sea herring, which now grows faster than it did in the 1930s, probably as a result of its presently reduced abundance. However, because the standing crop is now low, the predicted equilibrium yield to the herring fishery is considerably less than maximal.

A fish such as herring, then, functions in our eyes as an efficient prey, if it is fished to just such an extent that it gives the greatest sustainable yield to the fishery. In this way it is converting the greatest amount of zooplankton into potential human food. In the case of the North Sea herring fishery it was estimated that, prior to the recent decline in stock abundance, the maximum sustainable yield was about 750,000 tons a year. Using the same conversion factors as before, this yield of 0·2 gram of herring carbon per square metre per year represents about 20% of the primary production in the North Sea. Given the basic limitation imposed

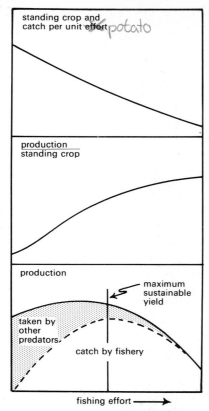

Figure 1.11 Fishing effort and its effects on equilibrium catch, production and standing crop.

by ecological efficiencies of 10%, such a pattern of fishing must be considered good harvesting when it is remembered that the rest of the primary production has to support the other commercial fisheries, not to mention invertebrate carnivores and seabirds.

Recruitment
Fisheries scientists usually define a *stock* as consisting only of fish old enough to be caught, and thus they separate *recruitment*, the addition of young fish, from the growth of existing members of a stock. The reasons for this distinction are that most fisheries data come from commercial landings, which do not include young or larval fish, and that larval and young fish often have feeding habits and living areas completely different from their parents. From the distinction comes the

following model, which is the basis of most of those used in fisheries regulation:

$$S_2 = S_1 + (A+G) - (C+M) \tag{2}$$

S_1 and S_2 refer to the standing crop of fish belonging to a stock at the beginning and at the end of a year. The gains to the stock are growth (G) and recruitment (A), and the losses are those due to fishing (C) and natural (M) mortalities. In an equilibrium fishery

$$S_2 = S_1 \text{ and so } (A+G) = (C+M).$$

A *year class* refers to the fish spawned in a particular year that survive to join the exploited stock at some later time. As we have seen, the survival of larval herring varies greatly from year to year, and thus the number of recruits is also very variable. Herring can be aged by an examination of the growth rings in their scales, and when this is done for large numbers of fish, usually taken from commercial landings, it is found that certain year classes dominate the catch. Such good-year classes can be recognized from one year's landings to the next, as in the example illustrated by figure 1.12. This refers to part of the North Sea herring, and it can be seen that the classes spawned in 1957, 1961 and 1964, and first making significant contributions to the fishery in 1959, 1963 and 1966, made up a large proportion of the catches between 1960 and 1967.

A major problem in estimating maximum sustainable yield is thus to determine the average rate of recruitment, the factors that control this average rate, and the factors that make for particularly good or bad-year classes. Although a relation might be expected between year-class strength and numbers of spawning adults, many fish stocks show little evidence of this. Indeed, one of the most widely used fisheries models assumes that the average rate of recruitment to a stock is independent of stock size. Thus, while predicting that overfishing will reduce yield, it does not suggest that it will have long-term deleterious effect on the stock.

This model seems to hold true for the plaice. In any year, the number of recruits appears to depend on factors unrelated to the size of the adult population, such as the amount of food available to the young, or the strength of the eastwards drift in the Southern Bight. If there are not too many bad years in a row, constancy of average recruitment is a factor that tends to stabilize plaice stocks, and thus the landings from the plaice fishery. Indeed, for many years landings of plaice from the North Sea varied relatively little, from 90,000 to 130,000 tons per annum.

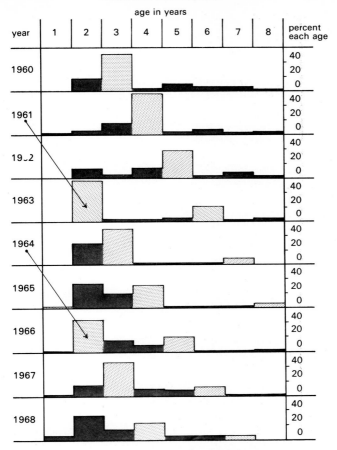

Figure 1.12 Year classes of North Sea herring.
Percentage age composition of fish landed by Scottish and other boats from the North-Western North Sea during the years 1960–1968. All fish have a nominal birthday on 1 January; thus fish of age 3 in 1960 were spawned at some time during 1957. Few fish of age 1 were landed in any year because most herring did not enter the fishery until they were 2 or 3 years old. Note the prominent year classes spawned in 1957, 1961 and 1964.

The story of the herring fisheries, however, demonstrates that a constant-recruitment model cannot always be applied. In this case there does seem to be a link between numbers of recruits and those of adult herring. This may be the result of herring laying fewer eggs than plaice, or of some other difference in the fishes' lives.

For various reasons it is not easy to assess either the absolute abundance of a herring stock or the absolute numbers recruited each year. If, however, the fishery is roughly in equilibrium and the average (although not necessarily the actual) recruitment is constant from year to year, it is possible to formulate models in terms of, and to attempt to maximize, the *yield per recruit*. This quantity is very convenient in fisheries regulation in that, given long-term constancy, it is independent of year-to-year fluctuations in recruitment. Except in fish farms, the growth rate of recruited fish is not under the control of the fisheries administrator; he can, however, legislate for the mesh size of nets used in fishing, and thus control the age at which fish become exploited; and he can attempt to reduce the total amount of fishing, thus reducing fishing mortality and allowing each recruit to live longer on average before it is caught. Thus the average size of landed herring may be increased, and graphs analogous to those in figure 1.11 may be drawn to predict the conditions that allow maximum yield per recruit.

Regulations of this sort were applied to the North Sea herring fishery quite successfully for a number of years. After reaching a peak of 1·4 million tons in 1965, however, landings decreased to half this in 1970–1972, and have continued to decline up to the time of writing. The explanation seems to be as follows.

From 1950–1962, and in spite of a gradual increase in fishing effort, North Sea herring landings were remarkably stable at an average of 850,000 tons a year. From 1963 onwards there was a big increase in effort that coincided with the recruitment of several good-year classes, in particular those of the 1961 and 1964 spawnings. These produced an increase in stock size sufficient to support the more intense fishery for several years, including the peak year 1965. However, surveys by research vessels showed a marked decline in the abundance of larvae, and by 1969 the numbers were only about 20% of those found at the beginning of the decade, indicating a reduction of about 80% in the stock of spawning herring.

Two other factors are also implicated in the decline in recruitment. These are an increase in the industrial fishery on the Bloden Ground for 1 and 2-year-old herring, and changes in climatic conditions that have resulted in changes in water temperatures and movements in the North Sea. These latter changes have affected the timing, extent and species composition of the spring zooplankton increase, with the result that biological conditions are now less favourable to larval herring than they were before.

It seems likely that the decline in the Peruvian anchoveta stems from similar causes: overfishing, that has reduced the stock of spawning fish; and also oceanographic changes in the upwelling area, resulting in decreased productivity of the plankton on which the anchoveta feeds, thus increasing larval mortality.

Problems of the fisheries

When the assumption of constant recruitment is correct, a fishery may be expected to be largely self-regulating. An increase in fishing effort may for a few years result in a larger total catch; but eventually this leads to a reduction in the abundance of the stock, and so to lower catch per unit effort. A new equilibrium is attained, with a catch that can be supported by the reduced stock.

Fishery effort is controlled by a mixture of influences: by national and international recommendations and regulations intended to maximize sustainable yields, and by the economics of fishing. Although the prices fetched by landed fish depend very much on the vagaries of supply and demand, it is on average possible to relate return on investment in a fishery to the catch per unit effort. Thus an underexploited stock, giving a good catch per unit effort, ensures a good return on capital invested in the fishery. This attracts further capital into the fishery, until decreases in catch per unit effort are such that the marginal return on additional investment is less than can be obtained elsewhere.

One of the problems facing the fisheries administrator is that the economic regulation of a fishery, as it has just been described, does not necessarily result in just that fishing effort necessary to give the greatest sustainable yield or the maximum yield per recruit. Since there is always a tendency to develop and use more effective fishing methods, or to exploit previously unfished parts of the stock, both of which result in short-term increases in catch per unit effort, most stocks are overfished; i.e. more fishing effort is employed than is necessary for the optimum yield, with the result that not only is catch per unit effort, and hence return on capital, lower than it might be, but the total catch is also less than the maximum possible. There is rarely any incentive, however, for the owner of a fishing boat or fleet to reduce his catch, however desirable this is in theory, for even if his reduction in effort is great enough to affect the stock significantly, the benefit is secured mainly by those other boats or fleets that continue to fish at their old intensity. The only solution is a general restriction of effort—but this is often an international, rather than a national, problem.

Overfishing of the kind described, which results in less than the maximal yield per recruit, or in a stock abundance lying to the right of the line of maximum sustainable yield shown in figure 1.11, is often called *growth overfishing* because the growth potential of recruits to the stock is less than maximally exploited. When recruitment is proportional to stock abundance, as in the case of the herring, there is also the likelihood that continued overfishing will cause serious and prolonged decline in the stock and yield of the fishery. This is called *recruitment overfishing* and demonstrates that the economic basis of fishery regulation discussed above makes sense only in the short term; for, in the long term, a drastic decline in yield means that boats must be laid up and capital investment under-used. A similar result may come about where the invention of a new type of fishing gear produces a temporary increase in catch:effort ratio, stimulating more investment than the fishery can support in the long term. In general there will always be a lag between increased effort and reduced catch per unit effort, and so we realize that only exploitation that is ecologically correct can be compatible with securing the best long-term return on capital invested in a fishery.

The problem is, however, that ecologically correct regulation requires more information than is available at the time that decisions must be made concerning fishing effort. This is because the data required by the technical models of fisheries, which are in themselves generally adequate for the task of prediction, come mainly from commercial landings. Thus, even in the best of cases, fisheries scientists can do little more than keep up with the existing state of affairs. Although usually able to explain an occurrence like the decline of the North Sea herring after it has begun to occur, the scientist is unlikely to have sufficient confidence in predictions made several years in advance to be able to convince his national or international superiors that severe restrictions should immediately be imposed on a fishery. The administrator, faced with the conflicting demands of the economics and the international politics of the fisheries—indeed of the livelihood of fishermen and solvency of ship-owners—is thus rarely willing to impose restrictions just in case they are necessary. By the time that the need is clearly apparent, it is often too late to save more than a part of the fishery.

Thus in the case of the North Sea herring, regulation has been carried out by means of quotas, but the way in which it has been necessary to reduce these, from the 500,000 tons per annum proposed in 1971 to the less than 250,000 tons fixed for 1975, shows the depletion of the stock and the inadequacy of regulation. It has already been mentioned that, if conservation were effective, the sustainable catch from the North Sea would be about 750,000 tons per year.

Because of the slowness of international action, a number of countries have felt it necessary to impose their own restrictions, often by the extension of territorial waters, as in the case of Iceland and as has been proposed for the protection of the Scottish West Coast herring fishery. This is an effective solution if it brings the major part of a stock under the control of a single authority, and if national fishing effort is suitably controlled; but it should be clear that in the case of fish like the North Sea herring and plaice, which breed in one place, feed when young in another, and as adults in a third region, only international action can achieve the necessary regulation of effort.

The crux of the scientific problem is the inability to predict far enough into the future on the basis of data from commercial landings which are mainly of adult fish. Attempts to find a way round this have involved surveys of larval fish abundance, in particular of herring in order to predict recruitment several years in advance. These surveys, however, are expensive in terms of research vessel time, and do not yet supply sufficiently precise information, even for the herring. The alternative is to attempt to predict larval survival, and hence recruitment, from observations on meteorological conditions over, and physical conditions in, the coastal waters which support exploited fish stocks. Such attempts require accurate models of the relations amongst physical conditions, primary production, zooplankton abundance, and larval survival, and unfortunately such models do not yet exist. It is a major task for marine ecology to develop these models.

Future exploitation of marine production

The fisheries

There seems little hope of increasing the landings of the conventional fisheries in many regions of the world. Some areas remain underexploited; the greatest remaining potential probably lies in the South-West Atlantic, the North-West Indian Ocean and the West-Central Pacific, but it seems unlikely that any of these regions approaches the productive capacity of the areas now most heavily exploited. It is thus improbable that future increases in fishery effort will more than double present landings, a result in agreement with theoretical predictions of production at the trophic level of exploited fish. Even to do this will require very careful regulation of existing and new fisheries. There is a danger that the effects of overfishing will more than counteract the benefits obtained from new fisheries.

If then there is only limited promise for the conventional fisheries, is there hope of harvesting other forms of marine production? Effective exploitation requires that the harvested organisms occur in concentrations sufficient to be worth catching. It is rarely likely to be possible, directly to fish phytoplankton or herbivorous zooplankton, because even in regions of high production their concentration per cubic metre is too low. It is more efficient, even in these regions, to catch fish like the anchoveta and herring that have carried out the work for us. On the high seas, only large high-value fish like tuna are worth catching and, being second or third-level carnivores, these animals are harvested at the expense of a great loss of potential yield.

Bearing in mind this principle, two alternatives present themselves for new and unconventional fisheries.

The first is exploit unusual animals at the top of the food chain, e.g. certain deep-water fish related to the cod. Although to do this requires a better understanding of the biology of these fish and the development of new deep-water fishing gear, it has been estimated that an annual world yield of 15 million tons might be realized.

The second alternative is to take account of the shoaling behaviour of some zooplanktonic crustaceans and fish. Thus it may be possible to use midwater trawls to catch the lantern-fish, euphausiids and shrimps that congregate in the "scattering layers" at depths of several hundred metres. The krill of the Southern Ocean are a further possibility, and one that the Russians have been investigating for some years. Although the total productivity of the region of the Antarctic Convergence is probably not so high as was once thought, there are areas of high production during the summer, and the phytoplankton here support swarms of these euphausiids. Estimates of yield to a fishery range from 10 to 50 million tons a year.

The development of suitable gear and methods may thus allow a harvest of new fish, squid and crustaceans at least equal in amount to the present landings of the conventional fisheries. It is probable, however, that much of this yield will not be consumed directly by human beings but, like about a third of present landings, will be used for feeding livestock and will thus play only an indirect part in our diet.

Fish-meal fisheries receive considerable criticism at a time when many people in the world are suffering from protein shortage, but it seems that in many cases the choice lies between fishing for meal and not fishing at all. Economic factors including, in particular in developing countries, the costs of transport and preservation relative to the price the consumer can afford, are partially responsible for this. The fish taken for meal are mostly small plankton feeders, like the anchoveta and young herring, and their conversion into cheap chicken may be a more efficient way of exploiting the marine food chain than by fishing the predators of the fish.

Mariculture

All the means discussed so far are essentially primitive methods of exploiting marine production. Is it possible that the development of marine husbandry will allow as great an increase in food production as did the change from the hunting economies of the Palaeolithic to the settled agricultural life-styles of the New Stone Age?

A hunter is little different from—and is in competition with—many other predators, and his prey is subject to many factors over which he has no control. A farmer tries to simplify the relevant parts of a food chain and to control external and internal factors affecting the growth and reproduction of his herds and crops, thus ensuring for himself a constant high proportion of the production on his land.

Mariculture, like terrestrial husbandry, ranges from casual attempts to improve the yield of wild or semi-wild unenclosed stocks, to strict management of caged populations of animals chosen for their efficiency in converting low-grade feedstuffs into high-grade fish or shellfish flesh. Maricultural practices are influenced by prevailing economic factors and popular views on the palatability and worth of various kinds of fish. In developing countries, fish-farming tends to be relatively labour-intensive and to exploit animals feeding on natural foods near the base of the food chain, thus producing amounts of protein that may contribute significantly to local diet. In developed countries, aquaculture tends to be capital-intensive, to use artificial feedstuffs and to produce relatively small amounts of high-quality high-price fish.

In the developed countries, the beginnings of mariculture may be traced to measures aimed at the conservation of existing fish stocks. With the expansion of trawling in the last half of the nineteenth century there came fears of overfishing. The marine hatcheries movement that began in the 1870s was stimulated by the success obtained with stocking of freshwater fish and rested on the false assumption that the recruitment of demersal fish was dependent upon the number of fertilized eggs produced by adults of the preceding generation. It was therefore held that depletion of stocks by fishing could be compensated by the release of large numbers of newly-hatched fry.

Evidence for the success of these operations in augmenting fish stocks was inconclusive, however, and further research meanwhile led to a greater understanding of the biology of exploited fish. The evidence obtained between the two World Wars was that recruitment was almost always independent of stock abundance, and so ways other than the addition of very young larvae were sought to increase fisheries yield. For example, it was known that the rate of growth of plaice in different parts of the North Sea varied according to food availability, and experiments were therefore carried out in which young plaice were transplanted from the crowded nursery areas of the Dutch coast to the relatively sparsely occupied Dogger Bank.

Next came proposals for the farming of enclosed populations of fish. This seemed more logical than attempts at the general improvements of a wild fish stock. It was argued that if each of the 35 million plaice caught by British trawlers in 1961 were given hypothetical allowance of 0·1 square metre of bottom, they could all be housed in shallow ponds covering 3·5 square kilometres. Plaice are tolerant of crowding and could, it was suggested, be reared on cheap manufactured food based on fish offal or agricultural by-products. While reaping the stock is the expensive part of traditional fishing, it becomes very cheap with pond-held fish: just a case of pulling out the plug and draining the pond. Because of variable supplies from the traditional fisheries, processing plant is generally inefficiently operated. Fish-farming would make for continuous supplies and thus ensure better use of this investment. In addition, supply could be better geared to demand, thus stabilizing prices.

Farms of this sort were intended ultimately to produce large quantities of cheap fish. Indeed, in South-East Asia fish ponds work well in this way, and given more research it seems likely that solutions could be found for the ecological problems of pond-rearing of flat fish that were discovered as a result of experimental enclosures of small Scottish sea-lochs during and after the Second World War.

However, a new pattern of marine fish-farming has appeared during the last decade, and seems likely to persist for a number of years. It is one of highly-managed farms producing small amounts of high-value fish. Perhaps the most successful operations are those involving salmon (and, more experimentally, sea-trout) grown on artificial diets made mainly from fish meal or locally-caught poor-quality ("trash") fish. Fresh-water farming of salmonids is of long standing, having originally been developed for stocking purposes and more recently adapted to produce marketable fish; but one of the main problems of these fresh-water farms is in obtaining an adequate supply of pure water to support the respiration of fish that must be grown in close proximity to reduce capital and running costs. Salmonids are very sensitive to reductions in the oxygen content of the water. The present marine farms have the advantage that the salmon and sea-trout can be housed in floating cages ventilated by tidal currents in the sheltered waters of a sea-loch.

Farms of this sort rely on taking young fish and rearing them to marketable size. The "recruitment" problem is solved in hatcheries, where the young fish are grown from eggs on a specialized diet of locally grown phytoplankton and zooplankton.

Flat-fish farming has so far not become an economically viable proposition, not because of the biological problems involved in rearing larval and adult fish, which have mostly been solved, but simply because it costs more to grow a plaice to marketable size than the fish will fetch when sold. Salmonid farming is successful, not only because the rearing of the larvae is comparatively well-understood, but also because the fish fetch more per pound than do plaice.

Many of the biological problems involved in fish-farming relate to the physiology or behaviour of the cultivated species. Fish are adapted to their life in the sea and not necessarily to the conditions of cultivation. Thus the body shape and feeding habits of plaice are indicative of life on the sea-bottom, and North Sea stocks of this fish have breeding behaviour well adapted to the local pattern of water movements, so that for example, larvae hatched in the Southern Bight metamorphose in time to settle on the Dutch coast. Although some of the adaptations of the plaice, e.g. its tendency to congregate on good feeding grounds, also fit it for life in cultivation, others, e,g. its delicate planktonic larva, do not. In the sea there occurs *natural selection* of those plaice best adapted to demersal life; this can be replaced by *artificial selection* of fish with characteristics considered most suitable in cultivation. This is a long-term process, requiring the breeding and careful selection of many generations of fish and is only now beginning.

The farming of salmon and sea-trout may be seen as a spatial dislocation of an existing marine food chain. Instead of fishing for salmon in conventional fashion, it is more efficient to harvest (often as a by-product of fishing for human consumption) fish at a lower trophic level and allow salmon to convert these to salmon flesh under the controlled conditions of a farm. The local environmental demands made by these farms are thus restricted to a supply of oxygenated water. Unless oxygen is obtained entirely from the air, which requires considerable mixing of the water body, there is an indirect link with primary production in terms of oxygen replenishment by local plant photosynthesis.

Oysters and mussels, which have also been successfully cultivated, feed directly on phytoplankton carried past them by water currents, and thus in regions of tidal movement can exploit the primary production of an area considerably greater than that immediately surrounding them. Because of their sedentary habit, they are potentially better converters of food than are active fish. Notwithstanding these great potential advantages, which imply a large return for a small capital investment, oyster cultivation is limited at present by the availability of sites with suitable water movements but without the great variations in salinity which occur in some estuaries. Other problems include sensitivity to pollution, which is an increasing feature of many estuaries.

The mariculture of the developing world provides a great contrast to the successful fish-farm practices of the developed countries. Although a great variety of fresh-water fish are grown in confinement in many countries, the practical culture of marine and brackish waters in the developing world is largely restricted to South-East Asia, where milkfish and mullets have long been raised in coastal ponds. These estuarine fish of tropical waters are tolerant of variable salinities and other environmental stresses, and feed near the bottom of the food chain. All these are desirable features for a cultivated fish, and the only major obstacle to the culture of milkfish and mullets is that it is difficult to induce them to spawn in captivity; thus, fish farmers rely on purchasing fry collected from inshore waters.

In Java, milkfish are raised in shallow ponds, called *tambaks*, that are connected to the sea by tidal streams or artificial ditches, and are built by clearing and excavating mangrove swamps. Because of their origin, these ponds are very fertile and the nutrient-rich mud of their shallow bottoms rapidly develops a cover of blue-green algae, diatoms, bacteria, Protozoa and other small animals. The milkfish, introduced as fry, feed on this productive benthic complex and grow rapidly, so that good ponds may annually yield several hundred kilograms of fish per hectare. In addition, extraneous fish and crustaceans, introduced with the seawater, also grow well and provide an additional harvest. It has been suggested that systematic elaboration of this *polyculture*, along the lines practised by the Chinese in their fresh-water carp ponds, might considerably increase total yields.

Nearly 200,000 tons of milkfish are raised annually in ponds in Indonesia, the Philippines and Taiwan. It has been estimated that there are about 4 million square kilometres of coastal wetlands in the world, mostly mangrove swamps in the tropics. Although for ecological reasons it would be unwise to sacrifice more than a small proportion of these to fish-farming, conversion of 5% could result in an extra production of several million tons of fish a year, using only the simplest of the techniques presently practised in South-East Asia. Improved methods, not more complex than those used by the Chinese in their carp ponds, could increase the yield to 10 or 20 million tons a year. Although it would probably be difficult to obtain sufficient wild fry at this level of production, it should be possible to develop strains of milkfish and mullet that will breed in captivity. There is thus considerable potential for the expansion of mariculture in tropical countries with low-lying shorelines.

Conclusion

It has been the intention of this chapter to demonstrate the link between the primary production of plants and the actual and potential yield of the sea to humankind. Because of the limitations imposed by ecological efficiences of about 10%, an oceanic primary production of 2×10^{16} grams of carbon a year implies that the maximum amount of fish obtainable by conventional fisheries is about one hundred million tons a year.

For a long time to come, these fisheries, which in 1973 landed 52 million tons, will be the main way in which we make use of marine production. It is therefore very important that fishing efforts be regulated so as to ensure the greatest long-term sustainable yield from exploited fish stocks. New and unconventional fisheries may one day supply several tens of millions of tons of fish, squid or crustaceans, but it is likely to be impracticable to attempt a direct harvest of primary production or, indeed, of any plankton except under the most favourable circumstances.

Fish-farming in the West will probably be limited to the production of small amounts of luxury fish, unless there is a significant change in economic conditions. In the under-developed world, however, there is considerable potential for increasing protein supplies by fish-farming coastal wetlands. Finally, it is worth repeating that coastal areas, which are often of great importance in the life of commercially important fish, and for various kinds of mariculture, are very sensitive to the effects of human activity.

FURTHER READING

Bardach, J. E., Ryther, J. H. and McLarney, W. O. (1972), *Aquaculture: the Farming and Husbandry of Freshwater and Marine Organisms*, Wiley-Interscience, London and New York.
    Contains a useful general introduction and 42 chapters (868 pages) on the culture of particular kinds of fish, molluscs, crustaceans and marine plants. Based to a large extent on interviews rather than on published scientific literature. Up-to-date and comprehensive, but lacking critical appraisal.

Cushing, D. (1975), *Marine Ecology and Fisheries*, Cambridge University Press.
    Up-to-date account of ecological aspects of fisheries at advanced level, by British fisheries expert. Includes chapters on primary and secondary production.

Gulland, J. A. (1971), "Ecological Aspects of Fishery Research." *Advances in Ecological Research*, **7**, 115–176.
    A good account by an FAO fisheries expert of some of the ecological and economic principles of the fisheries. Includes some mathematical treatment and numerous references.

Hardy, A. C. (1956 and 1959), *The Open Sea: Its Natural History*.
Part 1: *The World of Plankton*.
Part 2: *Fish and Fisheries*.
New Naturalist Series, Collins, London.
    Written by a distinguished and experienced marine biologist and beautifully illustrated. Highly recommended as an introduction to marine biology and the fisheries, but dated in parts. Part 1 now published as a Fontana paperback.

Kelly, M. G. and McGrath, J. C. (1975), *Biology: Evolution and Adaptation to the Environment*, Houghton Mifflin, Boston and London.
    An introductory text written from the same ecological perspective as that of the present chapter. Part 3 (pp. 253–471) provides a fuller explanation of many ecological points.

Raymont, J. E. G. (1966), "The Production of Marine Plankton," *Advances in Ecological Research*, **3**, 117–205.
    At the time of writing, this remains the most complete and convenient short review of primary and secondary production, but contains little synthesis of the information presented.

Riley, J. P. and Chester, R. (1971), *Introduction to Marine Chemistry*, Academic Press, London and New York.
    Chapter 3 gives an account of physical oceanography; chapter 7 reviews primary and secondary production and gives many references.

CHAPTER TWO

BIOLOGICAL CONSEQUENCES OF OIL SPILLS

A. NELSON-SMITH

Causes of marine oil pollution

The development and nature of the oil industry
Any explanation of the oil pollution problem, the forms which it takes, and the pattern of its occurrence along shipping routes or around seacoasts must begin with an outline of the development and nature of the oil industry itself. Petroleum became readily available about 100 years ago and has been in widespread use only within the present century but, even at the height of the recent so-called energy crisis, it was being shipped in greater quantities and larger carriers than any other product of commerce or industry. It was at first used solely as a source of kerosine (paraffin) for domestic lighting, to replace the already dwindling supply of whale oils and, by 1939, also provided the fuel for most forms of transport. The usual practice was to refine crude oil at its source, shipping the products to their users in tankers which, by modern standards, were very small.

During the Second World War, the dependence of the major industrial nations upon a regular supply of these products was made clear; subsequent political upheavals in many oil-producing countries as they themselves struggled to gain greater independence demonstrated the advisability of user nations acquiring a refining capacity of their own; thus they came to import crude rather than refined oils. Consumption grew by leaps and bounds as oil began to be used instead of coal in electricity generation and as a basic raw material of the chemical industry, particularly when the post-war demand for electrical appliances and synthetic plastics of all sorts accelerated dramatically. Both the nature and scale of tanker shipping has thus altered in a way that increases the incidence of serious pollution, since most refined products tend to be less troublesome than the raw material when spilt in open waters.

The possibility of spillage
In Europe, at least until offshore stocks can be fully exploited, consumption exceeds production by about the same extent that production exceeds consumption in the Middle East and North Africa. They are about equally balanced in both the Soviet bloc and North America, although the United States now imports in order to conserve her own stocks. Caribbean producers export their surplus, mainly to Europe, while in eastern Asia (especially Japan, which has a large demand but no domestic source of supply) the deficit is made up largely from the Middle East. The likelihood of spillage as a result of serious accident to a tanker, or even during routine operations, is obviously greater where the traffic is heavy; the heaviest traffic of all passed to the ports of north-western Europe through the Suez Canal and the Mediterranean until 1967 and then, until the re-opening of the Canal, around the Cape of Good Hope. Vessels are most likely to collide in the approaches to busy ports, particularly in the English Channel where shipping is denser than anywhere else in the world; they become stranded in narrow shallow waters—again the English Channel and southern North Sea, or the Bali-Lombok Strait on the tanker route to Japan. Occasionally even large modern ships in good condition have foundered or broken up in the storms of the South Atlantic or Indian oceans.

The greatest pollution potential arises from a major accident to one of the large fleet of VLCCs (very large crude carriers) of 200,000–300,000 deadweight tons, in which a single tank contains more oil than an entire pre-war tanker—although, as virtually all sea-going vessels now burn oil which is stored in bow tanks or the cavity of their double bottom, any collision or grounding carries the risk of some spillage. Offshore oil production is an important potential source of spills; although routine operations are carefully controlled, accidents can give rise to the release of very large volumes of oil which, because it passes through water immediately on its release, is a rich source of the lighter, more soluble and generally more toxic constituents which are removed during refining and may not even be loaded into crude tankers.

The discharge of ballast
Because the carriage of crude oil is a one-way operation—and "black" oils are anyway incompatible with products—tankers in this trade return empty of cargo; yet they need ballasting in order not to ride unmanageably

seawater on the return trip—which creates a problem, because up to one ton of oil in every 250 carried remains behind after unloading, clinging to the walls and internal braces of the tanks, but is readily washed off into the ballast-water. Even an Arabian loading-port, with rather relaxed standards of water quality, would find the pollution resulting from the discharge of such oily ballast unacceptable—but the alternative of pumping it to proper reception facilities on shore would be too expensive in turn-around time alone.

Ballasting therefore has to be carried out in two stages. On departure from the unloading port, water is pumped into some of the cargo-tanks; others are thoroughly washed *en route* and filled with a fresh load of seawater, clean enough to be emptied without further treatment as the next oil-cargo is taken on. Until the early 1960s, it was normal for both the original dirty ballast and the wash-water to be discharged to sea, once the tanker was clear of land. Many operators genuinely believed that the oily residues—scores of tons from a moderately large vessel—somehow "disappeared" in the vastness of the oceans although, in fact, such "operational" spillage was the cause of most minor pollution of bathing beaches and the steadily rising toll of seabirds, about which complaints had been made since the 1920s.

Between 1962 and 1964, a few of the major oil companies introduced the Load On Top (LOT) system in which, essentially, all oily water is passed by stages through a "slop-tank" where the oil floats to the surface: as it separates, relatively clean water is pumped away from underneath, leaving a large volume of recovered oil, on top of which the next cargo is loaded (hence the name). Most operators now use LOT procedures, although they are inapplicable in a few special circumstances. It has also been recognized that some tank-washing was, anyway, unnecessary; the larger modern tankers carry most of their water-ballast in separate tanks which never contain oil, and the mechanical or chemical processes used to separate oil from water are steadily being improved.

In these ways, it is claimed that operational losses have been reduced to only 10–20% of what they might have been; probably a disproportionately large share of this type of pollution arises these days from "rogue" tankers owned by the smaller less-scrupulous companies, while some continues to originate from general-cargo vessels, which also carry ballast-water in empty fuel tanks. International legislation still permits the discharge of oil to sea outside the prohibited zones which have been established around coastlines and in shallow enclosed seas, although some nations (notably Britain) forbid vessels flying their flag to do so, and many large operators have signed the voluntary Clean Seas Code which also strictly regulates polluting operations.

The amount of oil reaching the sea

There have been two recent attempts to review the numerous published estimates of the amounts of oil reaching the marine environment from various sources. The Massachusetts Institute of Technology, reporting on its *Study of Critical Environmental Problems* in 1970, suggested that a total of just over 2 million tons per year enters the sea from sources summarized in figure 2.1; a later report (1972) from the British Government's Warren

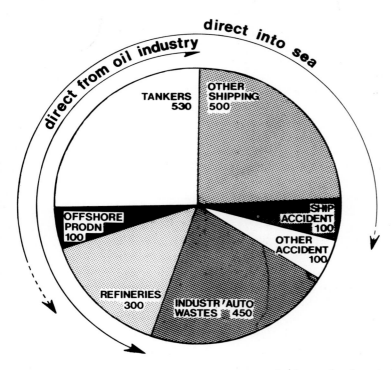

Figure 2.1 Sources from which spilt oil enters the sea (in thousands of tons); from Massachusetts Institute of Technology. *Study of Critical Environmental Problems*, 1970.

Spring Laboratory arrived at a total of 1·9 million tons, with only minor differences in the amounts assigned to each main source.

In both cases, the oil industry was thought to be directly responsible for rather under one-half the total, while about one-third was assigned to land-based sources. Of the latter, much arises from careless or thoughtless discharges from moderately large users (such as hospitals, hotels, plant nurseries or small factories) which nevertheless lack staff specially trained in handling oil, and the sum of many small spillages from motor-transport operation. "Do-it-yourself" motorists represent 60% of the market for lubricating oil, and each year in Britain alone pour about $1\frac{1}{4}$ million gallons (about $5\frac{1}{2}$ million litres) of used oil down the drain, most of which quickly reaches the sea.

The fall-out of airborne oil particles (whose special significance as potential carcinogens will be discussed later) and the release of hydrocarbons from growing plants or other natural sources both make a contribution which is considerable but very hard to quantify. It is obvious that spilt oil is also being steadily removed from the environment by various physical or biological agencies whose nature will be mentioned later in passing; again, it is very hard to quantify these processes.

Properties of spilt oil

Classification of oil pollutants
In the terminology of control legislation, marine oil pollutants are divided into persistent substances (crude, heavy fuel or lubricating oils and tank-residues) and non-persistent products (gasoline, kerosine and other light fuels, etc.). Generally, it is only the first group whose discharge is severely restricted although, of course, it is recognized that some of the lighter products can be highly dangerous when spilt in confined spaces or near regions of great human activity. Materials in the former group can travel great distances on the surface, where they are a menace to birds and a nuisance in many human operations, and will probably accumulate either on weather shores or at the bottom after sinking. It is, conversely, assumed that the latter group quickly become harmless by evaporation or dispersal in the water column. To a biologist, a more important distinction is between oils with purely mechanical effects and those which are also chemically active. Although the components divide in a broadly similar way, it is freshly-spilt crude oil and the lighter products which are most damaging, at least in the vicinity of a spillage.

Components of crude oil
Crude oil contains a vast number and variety of components, nearly all of which are hydrocarbons (i.e. they contain only carbon and hydrogen in various combinations) which form either the saturated and thus stable straight-chain branched or cyclic paraffins, or the unsaturated (i.e. chemically reactive) aromatics. "Cracked" products may also contain the unsaturated but non-cyclic olefins. The lowest in the paraffin series are gases (from methane to butane, cyclopropane and, under normal conditions, cyclobutane); these and the lower liquid paraffins, together with the

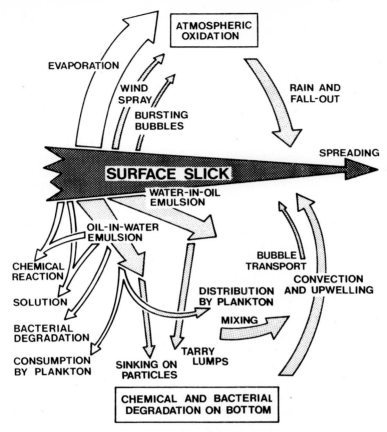

Figure 2.2 Processes which bring about the dispersal and degradation of oil spilt at sea (Nelson-Smith, 1972).

first few aromatics (from benzene to, perhaps, naphthalene) have a significant solubility in water.

When fresh crude oil is spilt at sea, it undergoes a variety of changes which modify its subsequent behaviour, toxicity and suitability for treatment by particular clean-up techniques (see figure 2.2). The lighter fractions, which tend to have the greatest toxicity, evaporate to the atmosphere or dissolve in the underlying water. If the surface is at all rough, droplets may be carried away both downwards into the water (where the incorporation of silt and other suspended particles may cause them to sink to the bottom) and upwards as an aerosol into the wind. The residue of the slick, forming a thinner film as it spreads, is likely to become stirred into a stable water-in-oil emulsion now widely referred to as "chocolate mousse"—for reasons which are obvious as soon as one has seen a freshly-deposited blob of this material. It can contain as much as 80% water while still behaving like thick oil, thus effectively extending the influence of a given spillage while making it very difficult to estimate the actual amount present.

Degradation of oil components

All oil components are subject to chemical degradation, whether in water or air, although at widely differing rates. Bacteria capable of breaking down hydrocarbons are present almost everywhere, although only in bottom sediments are they numerous enough to make much impression, and their activities are extremely slow in cold seas. The products of this degradation may be more widely utilized by other small organisms, while even quite large droplets of the more inert oils may be taken in and modified by some members of the plankton. In the short term, however, the impact of spilt oil on marine biota is much more notable than any influence which they may have on its removal.

Biological effects

External

All but the lightest, most volatile products can have a mechanical smothering effect. Large organisms immersed in the water may escape unscathed from a casual contact because they have surfaces which most oils will not wet, e.g. fish are covered with a mucus film, and seaweeds secrete a mucilage having similar properties. Seabirds are unfortunate in this respect, because their buoyancy and thermal insulation depend on feathers whose microstructure and waxy covering repel water but attract oil. This, if it is not too viscous, can rapidly penetrate the plumage, permitting the entry of water, while thicker residues stick to the feathers and weigh the bird down. Either way, the victim is almost certain to die from drowning, exposure or starvation.

Seabirds can keep up a normal body temperature (which is slightly higher than that of a mammal) as long as they have food reserves, to the extent of burning up most of their body muscles: American experiments have shown that heavily-oiled but well-fed ducks can survive exposure at temperatures as low as $-26°C$ for as long as 36 hours, but this says little about the prospects of an oiled bird under normal field conditions, unless it is rapidly caught and given skilled treatment.

Marine mammals are fewer and many of them (whales, sea cows and some seals) are virtually hairless; but fur seals, sea lions and sea otters, together with beavers, otters, muskrats and other species in fresh water, are affected in much the same way as birds when oil penetrates their fur. Seals, in particular, also seem very liable to blinding by freshly spilt oil. Smaller organisms, even though not readily wettable by oil droplets, may easily become caught up in larger masses, while sedentary animals and plants on the shore can do little to survive blanketing by a heavy stranding of oil.

As well as inhibiting movement and excluding oxygen, "black" oils cut out the light needed by algae and cause temperatures to rise rapidly in tide-pools and across polluted rock or sand surfaces during sunny periods when the tide is out. Such an effect can be fatal to corals in the tropics, where the temperature may already be near their limit of tolerance. A minor and rather academic point on the credit side is that some shore algae can function for longer when the tide is out if they have acquired a thin covering of one of the more inert oils, as this slows down the rate at which they become dehydrated in air. However, plants colonizing the upper shore and salt marshes, unlike these seaweeds, have a waxy covering which—as with birds—repels water but attracts oil. They also have breathing-pores which the thinner oils can penetrate.

Oiled foliage never recovers, so survival of affected plants depends on the season at which a spillage occurs and the extent of their food-stores. Pollution of a salt marsh during the winter looks bad but hardly affects the plants, which are more or less dormant; some perennials with large tap-roots can sprout again, several times if necessary, should it occur during the growing season, although annuals are usually eliminated by a single oiling.

Internal
Well-weathered residues, drifting across the oceans, become sufficiently innocuous to provide an anchorage for the gooseneck barnacles which normally attach to driftwood. Stranded residues, accumulating on rocks at the head of the shore, may also harden sufficiently for the settlement of acorn barnacles and are grazed from the surface without apparent ill-effects by limpets and similar molluscs. Mixed with sufficient mineral particles and organic debris, they may even help to build up a strandline which maritime plants can colonize. On the other hand, freshly spilt crude petroleum and many of its products can have poisonous effects of much greater severity than their mechanical impact. The low aromatics, typified by benzene, have great penetrating powers not only into whole organisms but also into the lipid (fatty) layer of cell membranes. It is thought that the mechanisms by which living cells regulate the entry or exit of active substances depend on the precise spacing of protein molecules on each side of this fatty layer; the intrusion of hydrocarbons into the lipid material

distorts the spacing and thus interferes with this control. The proper functioning of many nerves and—in higher animals—the brain itself similarly depends on fatty structures capable of being disrupted in this way. Some straight-chain and cyclo-paraffins are also noted for their effects on the nervous system, and are used as anaesthetics in medicine when, of course, their dosage can be carefully controlled. Disturbance of the proper spacing of lipid membranes may be a cause of the maldevelopment of eggs and young larvae which is often observed to follow an incident of oil pollution. Plants show some degree of recovery as these volatile hydrocarbons slowly evaporate, but only the simplest animals seem able to survive the damage which follows such disruption.

The fundamental effect of other toxic constituents seems to be upon enzymes or other chemicals vital to the regulation of internal processes, although their precise mechanism is still obscure. Again, some plants seem particularly resistant to these effects, notably the family of the Umbelliferae: water dropwort survives best of all on oiled saltmarshes, while a thin oil spray will kill weeds growing amongst rows of carrots (a close relative) without harming the crop. Non-hydrocarbon constituents—the so-called naphthenic acids and other compounds containing oxygen, sulphur or nitrogen as well as carbon and hydrogen—are more water-soluble than the lower hydrocarbons and are mostly more harmful. An incidental implication of this is that many oil/water separators used on shipboard, at coastal refineries or in industrial waste-water systems, are also functioning as extractors of poisonous constituents which are then usually discharged without further treatment. This is also a feature of separated ballast and wash-water discharged during conventional "Load On Top" procedures, although it is at least a lesser evil than the large quantities of oil which would otherwise be lost to sea.

Acute toxicity of oils

Median lethal concentration
Acute toxic effects are usually measured in terms of the concentration of toxicant which kills half the organisms in a test sample during some short period, usually 24–96 hours; as a small proportion of any sample of test animals or plants is bound to be unusually resistant, it would be unrepresentative to seek a total mortality. When testing drugs, food additives, etc., this provides a value known as the *median lethal dose* but, when aquatic organisms are placed in polluted water, it is not possible to control how much of the pollutant they actually take up so, here, it is more accurate to refer to a median lethal concentration (see, for example, figure 2.3).

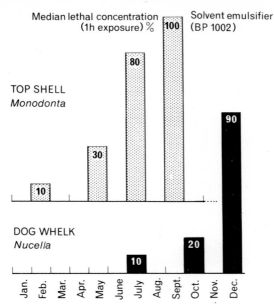

Figure 2.3 Sensitivity of two gastropod molluscs to an oil-spill solvent-emulsifier at various times of year, in aquarium tests during which animals were exposed for one hour and then transferred to clean water. The median lethal concentration is that which brings about the subsequent death of half the sample tested; animals were sensitive when the column is low, resistant when it is high. Summarized from data by Geoffrey Crapp (Oil Pollution Research Unit, Orielton, 1969).

Range of sensitivity
The different groups of marine organisms are variously constructed and, even within a single group, follow widely different modes of life; they thus show a great range from sensitivity to resistance.

Test organisms tend to be chosen for the convenience of laboratory staff and ease of maintenance in aquaria, rather than as adequately representing the wildlife of the region likely to be affected; and test conditions themselves differ very widely from laboratory to laboratory. References simply to the "toxicity" of any particular substance are therefore necessarily ludicrously oversimplified. Effects, even on a single individual, depend very much on its state of health, status within the habitat and past history: for example, tests on an early oil-spill dispersant whose marked toxicity was due mainly to the large amounts of aromatic hydrocarbon solvent which it contained, using the dog whelk *Nucella lapillus* and the top shell *Monodonta lineata* (both gastropod molluscs) showed that *Nucella* is far more sensitive in the summer, but *Monodonta* in the winter (figure 2.3). The former has a northern and the latter a southern distribution, although both occur in the British Isles, so—as might be expected—each is most resistant in the temperature range to which it is best adapted. During serious oil-spills in Milford Haven (very near its northern limit), *Monodonta* was the worst affected of the shore "snails" although, in western Cornwall (only 130 miles further south) it resisted *Torrey Canyon* oil and subsequent cleansing operations particularly well. This top shell grazes micro-algae from the rocks, while the dog whelk is a carnivore feeding on barnacles and mussels; so there are likely to be further differences in their respective susceptibilities in the field—especially in response to lower levels of pollution, whose influence is more insidious. Another gastropod tested, the winkle *Littorina obtusata*, also seemed to show a marked seasonal change in susceptibility but, this time, the sampling procedure was at fault; it was later realized that high mortalities recorded from spring samples had occurred because the population from which they were collected had eaten away all the seaweed available to it, and was thus starving.

Results of toxicity tests
It is additionally difficult to summarize the reported results of toxicity tests on crude oils since, although the amount of test material added to the water may be specified, the concentration of soluble constituents (in which, as has been pointed out, the main toxicity resides) is much more difficult to determine and is only rarely stated. According to the source of the oil, these may be present from about 4% to, exceptionally, nearly 40%. With these considerable reservations, the acute toxic response of the main divisions of marine life to typical crude oils can be generalized as follows. The micro-plants, larval and adult animals of the plankton, are rapidly killed within the range 2 ppm (parts per million) to 2%. Fish swimming in mid-water might succumb to between 10 ppm and 1%, while the crustaceans, molluscs and other bottom-living animals seem rather hardier, suffering lethal effects from 25 ppm to 5% concentrations.

In the more sophisticated tests, the aquarium water is circulated through the oil sample, but small droplets from it are trapped; more marked effects might be expected where, as in the field situation, particles are dispersed throughout the water, perhaps to be ingested or otherwise come into intimate contact with the organisms there. Many small marine animals, whether free-swimming or sedentary, feed by filtering particles from the water using net-like structures of interlocking limbs or gills. Selection is usually by size rather than taste or consistency, so that oil droplets falling within the size-range for which a particular filtering mechanism is adapted are likely either to be ingested or, if too noxious, to damage or inactivate the feeding structure itself. It is often supposed that, since most oil-slicks are observed at the surface, organisms at the bottom or in deep water can be in little danger from them; this ignores not only the many cases of oil sinking naturally, or being sunk deliberately during clean-up, but also its dispersion as droplets by the slow although powerful action of winds, waves and currents.

Many investigators, seeking to reproduce conditions on a heavily polluted shore, have exposed their test organisms to undiluted oil for shorter periods approximating to tidal exposure and recorded the subsequent mortality—which, in the case of a mid-shore winkle, for example, may range from 1% to 89% (averaging 53%) amongst 20 samples of crude oil from widely different sources. Most shore animals have the ability to close tightly against unfavourable conditions, so it is not surprising that, during exposures of an hour or so, followed by transfer to clean water, some snail-like gastropods, or bivalves with closely interlocking half-shells, show no mortality at all. The limpet, a gastropod which has no integral means of sealing its shell, seems unusually sensitive to petroleum and its products—with noteworthy ecological consequences which will be considered later.

Products themselves cause effects whose severity is as might be predicted from their content of light hydrocarbons. To typical organisms (i.e. neither extra-sensitive nor particularly resistant) gasolines have an acute toxicity between 60 ppm and 200 ppm; kerosine, gas-oil and Diesel fuels kill at 300–3000 ppm; heavier fuels and heating oils at 1000 ppm to 10% and lubricating oils at 3000 ppm to 20%. After short ($\frac{1}{2}$–6 hours) exposure of winkles to undiluted products, followed by washing and transfer to clean water, gasoline caused 80% mortality, kerosine and the crude oil from which it was distilled each 20–60%, but heavier products caused no immediate deaths. Some American No. 2 fuel oils are exceptional, containing over 45% low aromatics; as might be expected, these have caused devastating mortalities in the field, notable after the wreck of the tank-barge *Florida* near West Falmouth (Mass.) in 1969, which some commentators have unfortunately reported as though typical of marine oil-spills.

The comparable wreck of a coastal tanker carrying various fuel oils from the British port of Milford Haven caused negligible mortalities, the affected area being indistinguishable from those to either side as soon as the bulk of the spillage had been pumped or washed away. Of course, sublethal effects can rapidly lead to lethal consequences, e.g. gastropods anaesthetized by light products are likely to become dislodged from their chosen sites on intertidal rocks, after which they are unprotected from the normal hazards of dehydration, overheating, etc., or attack by such predators as birds when the tide is out, or fish when it returns.

Sublethal effects

Much lower concentrations will produce unfavourable consequences for communities or local populations, rather than individuals, which are necessarily never revealed by acute toxicity tests. The latter are the equivalent of a major oil-spill in confined waters, but marine organisms are probably exposed far more frequently to smaller quantities of crude oil, its products or aqueous extracts of either. Sublethal aquarium trials may suggest some effects of such exposure, but others may become apparent only in the field situation—when, of course, it may not be possible to lay the blame entirely or even at all on any particular incident or influence.

Plankton

Amongst the plankton, the division of algal cells is inhibited at oil concentrations as low as 0·01 ppm. Pollution at 0·02 ppm is sufficient to cause a slight inhibition of their photosynthetic processes (which are, ultimately, the support of all life in the open sea and which are also thought to make a major contribution to maintaining the oxygen level in the atmosphere).

 0·02 ppm is also an average value for the present hydrocarbon content of the water off the Atlantic coast of Canada—further south, in the Gulf of Mexico, three or four times this value has been recorded.

Eggs and larvae

In water containing 0·01 ppm, fish eggs hatch irregularly and late; larvae hatching from such eggs may be deformed, and the copepods which are their main food showed a delayed mortality to an 0·1% crude oil suspension after exposure for as little as five minutes. Abnormal development of young stages has been recorded at a level of about 1 ppm amongst lobsters, or between 10 and 100 ppm for the tougher barnacles and crabs; 10 ppm is about the lowest concentration at which an effluent can clearly be seen to contain oil. Larvae of sea urchins or starfish, which are highly sensitive to most forms of pollution, react in this way even to water stored over well-weathered tank sludge.

Effects on behaviour

Certain components of crude oil seem to act like a bad smell to some animals: salmon fry avoid concentrations as low as 1·5 ppm, which must at times adversely affect their migrations, while 10 ppm interferes with the ability of lobsters to find food. An emulsion of 0·1% Diesel and similar fuel oils inactivates the tube-feet of sea urchins, by which they cling to rocks or the kelp on which they feed. The same level brings about a rapid and complete inhibition of photosynthesis in the kelp plants themselves, while a 50% inhibition is brought about by only 10–100 ppm. Concentrations of crude oil over 1% stop the feeding of mussels, which are amongst the most resistant of animals inhabiting harbours and polluted estuaries. Lengthy or repeated exposure leaves them with insufficient reserves for growth or breeding, and may interfere with the production and development of the threads of tanned protein with which—like an echinoderm's tube-feet—they anchor themselves to the rock or quayside. Such delayed effects may escape the attention of short-term experimenters or the observers of a spillage in the field. The same irreversible damage to kelp physiology, mentioned above, may be caused by a similarly short exposure to only one-tenth the concentration of oils which produces an instant response—but it appears after a lag of seven days. Experimenters in a Scottish laboratory marked whelks (used in testing the toxicity of an aromatic-based oil-spill cleanser) by cutting a small notch at the mouth of the shell; animals exposed to concentrations so low that they had apparently been unaffected were returned to large live-cages but, months later, it was found that they had made insufficient growth to obliterate the notches.

Some aromatic hydrocarbons seem to have an almost specific chemical effect: as little as 0·001 ppm (one part per thousand million) has been found to block feeding and sexual behaviour in crabs (for which they rely on chemical "signals" in the water) for periods well beyond the end of the experiment.

Concentration of chemicals by oil
A rather more incidental effect also bothers marine scientists disturbed by the distribution of DDT and other persistent pesticides throughout the oceans: because these have a vastly greater affinity for oils than for water, it is feared that even tiny amounts of spilt oil, more or less harmless in themselves, could extract and concentrate such dangerous substances and act as the means by which they are introduced into food chains. No one has yet demonstrated that this actually occurs, but it is known that the planktonic crustaceans which form an important source of food for larger creatures as unrelated as herrings and whales readily incorporate petroleum hydrocarbons into their natural oil-stores. It should be noted that crustaceans, because they are closely allied to insects, are also highly sensitive to insecticides.

Wider implications

One possible result of the mortalities or sublethal effects which oil pollution exerts upon marine plants and animals is obvious and of direct economic importance—the loss of organisms utilized as human foods. Adult fish are usually thought to be able to avoid all but the most disastrous of spills, but their abundance and condition must clearly be affected to some extent by adverse influences on their eggs, young or food, and they cannot so readily avoid chronic widespread pollution at low levels. Fin-fisheries have declined all over the world but, although telling comparisons have been made with increases in the quantity or extent of various forms of pollution, it is often equally likely that over-fishing or natural changes are to blame; only rarely can any simple cause be identified as solely responsible.

Other organisms of commercial importance are less mobile or more selective and thus more vulnerable to a large spillage: various fisheries for oysters and other bivalves, together with seaweed beds off Japan and the Pacific coast of North America, have undoubtedly been ruined by particular incidents, while less well-substantiated claims have been made in respect of chronic pollution from, for example, oil production installations along the Louisiana coast. Locally, oyster production has even improved as a result of the greater sensitivity of their pests to low levels of contamination but, unfortunately, some hydrocarbons or oil derivatives have a strong unpleasant taste. Even where commercial stocks have suffered neither heavy mortalities nor detectable loss in condition, they may be rendered unsaleable because of an oily taint. 0·01 ppm gives rise to a marked taste of oil in oysters, finfish or edible seaweeds which may persist for several months, while even lower concentrations can readily be detected in drinking water. Seafood sales are very vulnerable to "scare" stories and may plummet even when there is no real reason, e.g. fish sales in the Paris market declined to one-quarter of their normal level after news of *Torrey Canyon* pollution, even though scarcely any of its supplies came from the affected area.

Ecological consequences

Disturbance of the ecosystem and succession
It is the constant message of ecologists that no species is without importance in preserving the "balance of life", even when its function is not well understood or appears superficially to be contrary to human interests. Its thoughtless or unintentional elimination carries the danger of subtle and probably unpredictable repercussions, well illustrated by the effects of serious oil pollution—particularly when followed by "cleansing" with damaging solvent-emulsifiers—on temperate rocky seashores. Under normal conditions, the severity of natural influences on this habitat can be judged by a glance at the organisms dominating the intertidal zone:

in shelter it is likely to be densely covered by the larger brown algae, but they are sparse in wave-swept situations, where the rocks are encrusted with barnacles and mussels. Green algae occur on open shores only in small isolated patches, but they dominate areas influenced by fresh water or enriching additions (where streams or small sewage outfalls cross the shore, or in small harbours). The herbivorous winkles or limpets are to be found on most rocky shores but, broadly speaking, winkles predominate in shelter and limpets are more numerous on exposed shores.

The particular sensitivity of limpets to hydrocarbon pollution has already been mentioned: after serious incidents not only in north-western Europe, but also in South Africa and elsewhere, they have been virtually eliminated from the worst affected stretches. Only then does it become apparent that the green seaweeds are excluded by their constant grazing over the rock surface, rather than because physical conditions are unsuitable for their settlement in the first place. Thus wave-beaten reefs and headlands have acquired a green covering which, to the experienced eye, seems ridiculously out of place. This "turf", in turn, can provide sufficient shelter for young brown algae to obtain a foothold; once established, these perennial plants may persist for several years, until the combined onslaught of recolonizing herbivores and winter storms removes them. During this time, existing barnacles are smothered by the encroaching seaweeds, and the settlement of further sedentary animals is severely impeded; this, in turn, affects the dog whelks, carnivorous worms and small fish whose food supply is thus diminished, although a few of the species associated with the weed may appear in their place.

Locally, the ecosystem has been disturbed at all levels, from the plankton which includes the youngest stages of both plants and animals to the large fish which prey on shore animals when the tide is in. An almost invariable effect is to reduce the variety of plants and animals, although after a single incident—even if catastrophic—a seashore previously in good condition should recover within three to seven years. However, each repetition of pollution during the recovery period will increase the likelihood of permanent damage. In Milford Haven—where the rapid dispersion of small spillages prevents most of them from reaching the shore and a widespread rich flora and fauna provide strong stocks for the rapid repair of local damage—a small flow of oily water has nevertheless created a permanent anomaly in the distribution of seaweeds and sensitive herbivores along a short section of the coastline (figure 2.4). Under less-favourable conditions (e.g. in parts of the tideless oil-port of Marseille) it may be difficult to find more than two or three of the toughest species on the quaysides. Such a reduction in the variety and—in these circum-

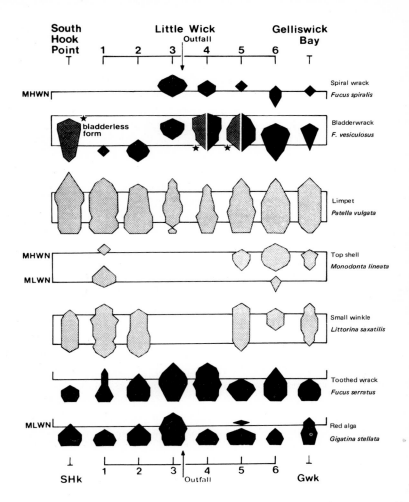

Figure 2.4 Influence of a refinery effluent on part of the north shore of Milford Haven. The densities of various seaweeds and invertebrate animals were recorded down intertidal transects at several stations to each side of the outfall, showing a reduction in grazers (stippled) and an increase in seaweed cover (black). At stations 3 and 4, the bladderwrack occurs partially in a form more typical of the exposed conditions at South Hook Point, which is unusual on unpolluted sheltered shores. Due to peculiarities in the abundance scale used, limpets at these two stations appear to be more numerous than they actually were. The level of high or low water of neap tides is given to provide reference points. From the author's chapter in *Petroleum and the Continental Shelf of NW Europe*, Vol. 2, Cole, 1975.

stances—probably also the total quantity of living material cannot be desirable by any standards.

Human amenity
Taking the narrower viewpoint of human amenity, those species most tolerant of chronic pollution are mostly the least useful or attractive. Variety and interest in natural habitats are now demanded not only by the student groups or amateur naturalists who are making increasing numbers of organized field visits to the countryside and seashore, but also by ordinary people taking individual recreation there. This is particularly true of seabirds, where popular interest combines with special susceptibility to oil pollution. Rocks on the shore, covered by soft green algae, are slippery and remain wet when the tide is out; organic rubbish such as picnic remains or stranded weed do not disappear so rapidly from the sands if the creatures which normally scavenge them have been killed off.

Saltmarshes commonly occur in the estuaries and sheltered inlets now favoured not only as sites for the unloading and refining of oil products but also for large industrial users such as electricity generating stations. Their vegetation, too, may be particularly susceptible to repeated oil pollutions. Such marshes are regarded by biologists as an important source of nutrients for young fish in their estuarine nurseries throughout the year, as well as for many migratory birds over winter; they may also provide free grazing during lean times for nearby farm stock, and they have a more general value as a pollution- and silt-trap. An extensive saltmarsh in Southampton Water has, in places, lost more than a metre's depth of soil since it became chronically polluted by oily effluents which destroyed the marshgrass whose roots normally bind it. Not only this eroded soil, but also the suspended silt which the marsh vegetation would have continued to trap, now add to the dredgers' load. Fortunately, with improvements in water quality some plants are now recolonizing the devastated area (figure 2.5).

Carcinogenesis

One biological consequence of oil pollution is worrying, partly because the extent of its impact has not yet been clearly established but mainly because, as far as humans are concerned, that impact is an insidious and long-term one for which there is no easy remedy. Crude oil contains a variety of known or suspected carcinogens, particularly the polynuclear aromatic hydrocarbons such as benz(a)pyrene, which is also an active constituent of tobacco smoke. Heating the oil to about 400°C or above (as in an oil-fired furnace or during refinery processing) increases the content of such

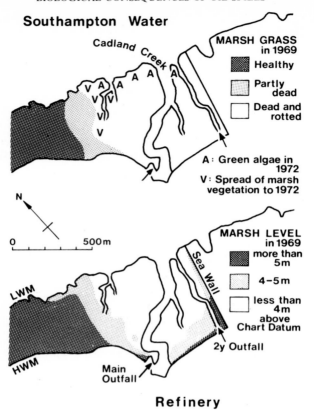

Figure 2.5 Influence of a refinery effluent on a saltmarsh in Southampton Water. The upper map shows the extent of damage to cordgrass *Spartina anglica*, which once covered the marsh, revealed by surveys made during 1969; the lower map indicates the drop in marsh level which resulted. After improvements in the quality of the effluent, marsh vegetation (V) and green algae (A) were spreading in 1972. (From the same source as figure 2.4.)

substances. It may be significant that fish caught whilst feeding around waste outfalls from petroleum-based industrial plant occasionally have tumour-like growths around the mouth; similar irregularities in growth have been reported amongst encrusting animals growing over jetty piles preserved with petroleum products. It is thus of some concern that as much as 2% of the oil fed to a ship's furnaces (not to mention oil-fired installations on land, motor vehicles and aircraft) emerges unburnt from the smokestack. As so much of Earth's surface is covered by ocean, the majority of this strongly heated airborne oil must fall back upon water rather than land.

Possible carcinogens are taken up and stored, just like other hydrocarbons, by bivalves and other invertebrate animals. Some workers have reported that they are rapidly lost on transfer to clean water (although obviously not if the animals remain in a polluted environment), while others claim that foreign hydrocarbons are retained for long periods, or even incorporated into the flesh of the mollusc. It certainly seems to have been established that mussels can deal much more readily with hydrocarbons of recent natural origin than with those from petroleum, and that they discharge aromatics slowest of all. However, fish can readily metabolize most hydrocarbons, excreting the breakdown products, so there appears to be little danger of their accumulation up food chains in the manner now well known to occur with chlorinated hydrocarbon pesticides. The slight changes in chemical structure which are likely to occur during their deposition in marine sediments or passage through living organisms may well render harmless these compounds which have such a specific action although, on the other hand, polynuclear aromatic hydrocarbons can be manufactured from related or simpler compounds by many marine bacteria and algae.

Some protection may be afforded by the sensitivity of the human palate to traces of certain hydrocarbons in food, but there are no grounds for assuming that all petroleum-derived carcinogens have a strong taste. In any case, some people do not find an oily taste repugnant; others may be unable to detect it, while the flavour imparted by a variety of culinary methods would anyway disguise it. While we continue to accept greater risks in other areas, there is little point in shunning seafoods on the grounds of possible carcinogenesis, but the part which oily wastes play in enhancing the hazard plainly requires much more investigation.

Treatment of oil spills

At all stages from the production of petroleum to the ultimate use or disposal of products derived from it, enormous improvements could be made in preventing spillage or providing proper facilities for waste treatment; but, for as long as oil is used in bulk, large quantities will continue to pollute the marine environment—if only as the result of human errors, carelessness or misunderstanding. It is important to know how best to deal with such spillages, bearing in mind that even the most rapid and efficient removal of the spilt material must leave behind small quantities of oil, together with a disproportionately large amount of leached water-soluble components while, at the other extreme, some methods may cause more biological damage than would have resulted from leaving the original pollution untreated.

The need to be prepared

All but the thickest oils spread quickly across the water surface, are dispersed further by winds and currents, and have components which rapidly become distributed into both the air and deeper layers of the water. The essence of any form of treatment is therefore speed of action, which can come only from advance planning. It is important to arrange both a working structure (which usually involves the emergency services such as fire-fighters and coastguard, together with organizations possessing the necessary plant and supplies) and a plan of action related to the nature of the area in question. A heavily-used and already much-polluted dock will present problems or offer possibilities which differ greatly from those of an inaccessible or rugged rocky coastline of particular biological value; between these extremes there may be beaches of some interest to biologists but which also support a valuable tourist trade. The best mode of treatment for each stretch of coast or offshore region, taking into consideration its topography, prevailing currents and so forth, as well as its particular use or value, must be firmly agreed amongst all interested parties well before any need arises to use the plan, or emergency operations may well bog down in arguments, recriminations or even legal action.

Methods of removal

Wherever possible, complete removal is the best treatment: this is easy on hard level ground, and there is a variety of devices for skimming, trapping or absorbing oil on the surface of enclosed still waters. On shores inaccessible to heavy plant, much can be done by crews with shovels and plastic or paper sacks, although some thought must, of course, be given to proper disposal of the material collected. Saltmarshes are a noteworthy exception—no treatment can remedy the damage already caused and most will only make it worse, so it may be best to leave the oil where it is unless there is an immediate risk to birds congregating there. In waves or strong currents, fixed spillbooms are inefficient at trapping floating oil and few skimmers function well. The BP "Vikoma" system is probably the best recovery method so far developed: to prevent further spread the slick is enclosed in a stable boom, deployed from a raft under remote control, from which it can be pumped by way of an elaborate arrangement of rotating discs which pick up oil because of its greater adhesion and thus take up less water than surface-skimmers (figure 2.6).

It is important that oil still at sea should be prevented from reaching the shore so, where recovery is impossible, second-best methods have to be used. These consist of sinking or dispersing the oil by adding further

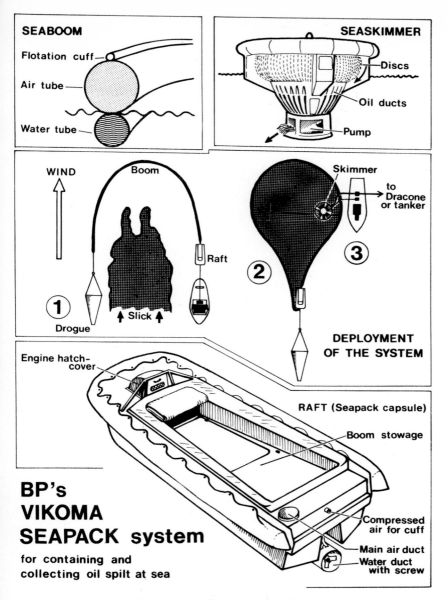

Figure 2.6 Operation of the BP "Vikoma" system of oil-spill containment and recovery. An oil slick drifts with the wind and would be drawn underneath a fixed boom. The "Vikoma" sea-boom has to be stable but flexible for open-sea conditions; it is deployed automatically (1) from a storage raft which at the same time inflates both the water and air tubes. When the slick has been enclosed (2), it can be allowed to drift while the skimmer (3) removes the oil to a container for disposal. The revolving discs of the skimmer pick up oil in preference to water; it is continuously scraped off into ducts feeding the pump. Moving parts are operated hydraulically from a service pack on the work-boat.

materials which may themselves have undesirable biological effects. There is still much disagreement about which method does the least damage; it probably depends on individual circumstances. Sinking can be carried out by adding inert materials such as specially-treated sand or pulverized fuel ash, and it carries the oil to the bottom, where the bacteria most capable of degrading it are concentrated; this can ruin the feeding or breeding grounds of fish and invertebrates, and may foul bottom-fishing gear.

Dispersion is achieved by spraying with solvent-emulsifiers and agitating the mixture vigorously; emulsification of the oil is brought about by carefully selected detergents whose penetration of the oil mass is speeded by the solvents. When properly carried out, this results in a cloud of droplets which should not re-coalesce; the theory is that various processes of degradation are hastened by thus increasing the surface area. However, detergents themselves are bio-active; increasing the surface area also enhances the extraction of water-soluble toxicants, and some droplets are bound to fall within the size-range favoured by filter-feeders. Early formulations contained highly toxic aromatic hydrocarbon solvents and were undoubtedly much more damaging than the oil itself, although "second-generation" products have been greatly improved both in efficiency (so that less needs to be used) and reduced toxicity.

Stranded oil can also be removed from hard surfaces by spraying with solvent-emulsifiers and subsequent thorough washing down with clean seawater, but this is best restricted to high-amenity or working areas such as slipways and quaysides, as it also enhances the biological damage by enabling the oil to "wet" or penetrate the outer covering of living organisms and by spreading it into cracks, crevices and throughout the depths of rock-pools. Most of the damage which followed the stranding of *Torrey Canyon* oil in West Cornwall was the result of improper and over-enthusiastic application of toxic so-called "detergents". Applied to sandy or pebble beaches, these merely facilitate the deeper penetration of the oil. Alternative treatments include the use of fine limestone chips, exfoliated mica granules, shredded plastic foam or other absorbents, and burning off the oil with the use of natural or synthetic "wicking" materials. Each is most appropriate to a particular set of conditions. The secret of success is to suit the clean-up method to the conditions prevailing at the time, while remembering that the ultimate aim should be to reduce damage to the marine environment as a whole, rather than merely to shift an undesirable problem from your beaches into some other authority's area as quickly and cheaply as possible.

FURTHER READING

General aspects of marine oil pollution
Nelson-Smith, A. (1972), *Oil Pollution and Marine Ecology*, Elek Scientific, London.
Smith, J. E. (ed.) (1968), *'Torrey Canyon' Pollution and Marine Life*, Cambridge University Press.
Hepple, P. (ed.) (1971), *Water Pollution by Oil*, Institute of Petroleum/Applied Science Publishers, Barking, Essex.

More specialized publications
Cowell, E. B. (ed.) (1971), *Ecological Effects of Oil Pollution on Littoral Communities*, Institute of Petroleum/Applied Science Publishers, Barking, Essex.
Cole, H. A. (ed.) (1975), *Petroleum and the Continental Shelf of North-West Europe*, Vol. 2, *Environmental Protection*, Institute of Petroleum/Applied Science Publishers, Barking, Essex.
Prevention and Control of Oil Spills (Proceedings of joint conferences with US Coast Guard and Environmental Protection Agency), American Petroleum Institute, Washington, D.C. Published in alternate years from 1969, and containing much information on biological effects.
Moulder, D. S. and Varley, A. (1971), *Bibliography on Marine and Estuarine Oil Pollution*, Marine Biological Association, Plymouth, Devon; supplement, 1975. Lists every major scientific publication on the subject.

Books discussing other forms of marine pollution
Ruivo, M. (ed.) (1973), *Marine Pollution and Sea Life*, FAO/Fishing News Books, West Byfleet, Surrey. Over 100 short papers on world-wide problems.
Moorcraft, C. (1972), *Must the Seas Die?*, Temple Smith, London. A short account, in paperback, which also discusses overfishing and other conservation problems.
Man's Impact on the Global Environment (Study of Critical Environmental Problems), MIT Press, London.

Useful periodicals (amongst many)
Marine Pollution Bulletin (monthly; Pergamon Press, Oxford). Carries up-to-date reports and comment.
Marine Pollution Research Titles (monthly; Marine Biological Association, Plymouth, Devon). Records details of all scientific publications filed by this Laboratory's Pollution Library.

CHAPTER THREE

INORGANIC WASTES

E. J. Perkins

Introduction

The oceans cover some 71% of the earth's surface to an average depth of 12,500 ft (3800 m). If the world were a perfect sphere, the entire globe would be covered to a depth of 8000 ft or 1300 fathoms (2500 m). Thus it is that the sea appears limitless to earthbound man, and naturally offers an attractive means of disposal for the dejecta of civilized life, as well as a source of some of the raw materials necessary for the continuance of that life.

The continental shelf
However, there are problems of crucial importance which these statistics do not reveal. Of first consequence is the fact that the productive continental-shelf areas, i.e. those waters with a depth of less than 100 fm (200 m), comprise only about 3% of the earth's surface. Even if we include the very rich waters of high latitude, and the areas of upwelling off the western coasts of the African and American continents, by far the greater proportion of all seawater is relatively unproductive, for much of it belongs to the impoverished central water masses. It follows then that most of the effective production of food from the sea comes from a rather limited area (and a still more limited volume) of which the continental-shelf areas, i.e. the neritic or coastal waters, are particularly exposed to the influence of human activity.

These waters are of great importance to the fishing industry, both as nursery grounds and as the sites of fisheries; but it is here that the direct impact of industry is also felt. It is on this shelf that all the submarine exploration for oil and much mineral and gravel extraction has so far been carried out; it is from here that cooling water is drawn by large power stations; it is to here that liquid effluents are released; and it is here that by far the greatest proportion of all dumping occurs. In addition, it is this area and its bounding coastline that is particularly subject to the influence of recreational activity (such as sea angling, yachting, water-skiing, walking and horse riding), dredging, land-fill reclamation schemes, and similar schemes dear to the real-estate developer. This, then, is a part of the biosphere in which a great conflict of interest occurs.

The inorganic wastes

The wastes and pollutants which the neritic waters may receive are varied (see Table 3.1). Of these wastes, sewage is by far the most important in terms of total input and, together with the other organic wastes—particularly the chlorinated hydrocarbons which include pesticides, vinyl chloride wastes and polychlorinated biphenyls—it has been a major preoccupation in recent times. In general, the inorganic wastes have been rather ignored, except in the cases of the trace metals (particularly mercury) and certain aspects of dumping. To a lesser extent, phosphates have received some attention in investigations of the problems created by marine eutrophication but, as will be shown below, some of the less-acceptable symptoms of eutrophication arise not simply from the amount of phosphate present in the water, but rather from other influences, some of which are biological, acting in conjunction with the phosphate present; eutrophication may reach unpleasant proportions, even though the phosphate concentration is in itself not remarkable.

Some aspects of inorganic waste disposal are not susceptible to a simple ecological investigation. The same hydrographic forces act upon wastes (organic and inorganic) and on the biota, and all three must often be considered together.

Bearing the above considerations in mind, there is little point in making a routine catalogue of effects which will probably contribute little to knowledge or understanding. It is intended therefore to review briefly the problem of inorganic-waste disposal in general, and to examine in greater detail cases from studies (not previously published) undertaken in the Solway Firth as a further contribution to our knowledge of the sea as a whole. This is intended to show to interested specialist and non-specialist readers the kinds of problems confronting workers in this field, and the methods by which such problems may be investigated.

The problem in general

The problems resulting from the disposal of inorganic wastes to the sea

Table 3.1 Sources of chemical pollution

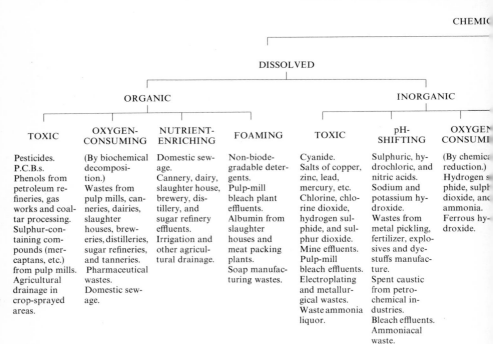

arise from three principal causes: (a) pH, (b) the major ions and radicles, and (c) the trace or heavy metals.

The composition of seawater is such that there is a slight excess of the strong cations Na^+, K^+ and Ca^{2+} over the strong anions Cl^- and SO_4^{2-}. In consequence, a considerable amount of carbon dioxide can be carried in solution as bicarbonate and carbonate, thus

$$\underset{\text{in air}}{CO_2} \rightleftharpoons \underset{\text{in water}}{CO_2} \rightleftharpoons H_2CO_3 \rightleftharpoons H^+ + HCO_3^- \rightleftharpoons 2H^+ + CO_3^{2-}$$

The resulting solution is slightly alkaline with pH between 8·1 in winter and 8·4 in summer, depending upon the relative amounts of CO_2 produced by respiration and consumed during photosynthesis. Thus any change in pH indicates a shift in the equilibrium and may be used to measure these activities in normal seawater. The system as a whole is weakly buffered; although it is often considered to be a strong buffer system, it is not sufficiently widely appreciated that in normal circumstances (let alone the classical cases of isolated basins and sunlit rock pools, in which the pH may

INORGANIC WASTES

ods of analysis (after Waldichuk, 1967).

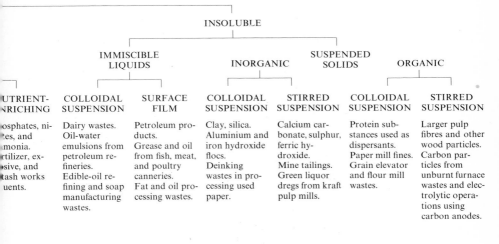

fall to 7·0 and rise to 9·6) diurnal as well as seasonal changes, due to entirely *normal* biological processes, may be evident. In addition, a flush from the land may lead to a depression of the pH, since in these cases salinity and pH are related.

Variations in pH
In daily records of water quality kept at the University of Strathclyde Marine Laboratory, Garelochhead, from 1967 to 1975, such changes were typical. However, in 1974, which had a colder summer than normal, dense phytoplankton blooms, particularly of diatoms, persisted from early April until mid-October: in this period the pH of the surface waters was measured on 88 days and, of these, 24 values (27%) exceeded the normal summer value of 8·4, and 6 values (7%) were 8·8 or more. These values, it should be appreciated, were obtained in the open waters of the Gare Loch. Persistent thick phytoplankton blooms had not been noted in the years previous to 1974, nor were they noted in 1975. However, the condition of 1974 was noted elsewhere in the Scottish west-coast sea-lochs, particu-

larly in Loch Sween. The strike of the Glasgow sewage workers, which began on October 7 and continued until December 2, 1974, had no obvious effect upon the phytoplankton of the Gare Loch, since the blooms did not persist after mid-October. Moreover, since the blooms of 1975 were rather poor (a situation also typical of Loch Sween) it is reasonable to conclude that the phenomena observed in 1974 were not a manifestation of eutrophication, but of the climatic and hydrographic conditions of that year, which were peculiarly favourable to the development of phytoplankton blooms.

Such changes mean that the normal buffering power of seawater is not an entirely dependable means of dealing with those effluents which have either a very high or a very low pH. This is more particularly true since, in general terms, the effective limits of pH for marine life are 6·5 and 9·0; in the case quoted for the Gare Loch, the pH noted approached a harmful level at times.

pH and toxicity
Without discussing in detail cases which have been dealt with elsewhere (Perkins, 1974), it is worth noting that pH itself is not only a limiting factor to marine life, but it may increase the effectiveness of a toxicant as the departure from the normal seasonal value becomes more pronounced: thus J. R. S. Gilchrist showed that a 15% solution of phosphogypsum effluent added to seawater produced a 50% mortality of the hermit crab *Eupagurus bernhardus* after 24 hours' exposure. In contrast, effluent which was first neutralized required a concentration of 37·9% to produce the same mortality, a result similar to that reported from fresh water by Howard and Walden (1965).

The dangers of using data from fresh-water situations
Too often workers in the marine field draw upon data from the more extensive work in fresh water. But there are serious pitfalls, for, while it is correct to use fresh-water data as a *guide* to what we may expect in the sea, data cannot be directly transposed. It is, for example, unreasonable to suppose that the dissociation of salts and the influence of pH are the same in media which have salinities of 0·030 and 35·0 g/kg respectively. On the other hand, we might reasonably expect that the solubility of most salts is less in seawater than in fresh water: but this is not so, for gypsum ($CaSO_4 \cdot 2H_2O$), which has a solubility of 0·2% in fresh water is roughly twice as soluble in seawater, depending on its salinity. This is a particularly interesting example, which is discussed in more detail below.

In a similar way it has been possible to show that the barnacles *Elminius modestus* and *Balanus balanoides*, diatoms *Melosira* sp., the green alga *Enteromorpha* sp., and the mullet *Mugil* either inhabit or, in the last case, frequently visit rocks bathed by seawater with local concentrations of fluoride of 2·1–40 ppm and a local pH which at times may fall below 6·0. This is in sharp contrast to the situation in fresh water where (according to *Water Quality Criteria*, 1972, commissioned by the US Environmental Protection Agency) a concentration of 2·3 mg/l (ppm) sodium fluoride (i.e. 1·04 ppm F^-), is lethal to rainbow trout *Salmo gairdneri*. *Water Quality Criteria* also discusses the toxicity of chlorine, widely used to control fouling growths in the culverts of power stations which use seawater for cooling purposes. It is unfortunate that such an authoritative document deals with this problem in terms of the fresh-water situation and outdated marine information, ignoring the well-publicized work of the British Central Electricity Generating Board and related bodies. The Report suggests the formation of chloramines and the possibility of persistent enhanced toxicity from this source. There is, however, no evidence to support this at the concentrations of chlorine used. Experiments designed to demonstrate chloramine formation in seawater have consistently given negative results.

Much has yet to be learned, not only of the chemistry of seawater, but of the differences in toxicity resulting from the differences in the chemical forms in which a particular element may be introduced into the seawater.

Turning now to inorganic chemicals as a whole, *Water Quality Criteria* defined those inorganic chemicals which exert an important influence on marine life (Table 3.2). These criteria are useful where an effluent of relatively simple composition is under consideration, but may not be of great value in regard to dumping, or in a situation where a marked precipitation is accompanied by incorporation in sediments, or where a chemical transformation of a component can occur.

The problem of dumping

The question of what happens when a waste has become incorporated in a sediment is of prime importance in relation to dumping. This problem arises largely from a hard core of very toxic pollutants which cannot easily be disposed of on the land or in fresh water, and from the disposal of large amounts of solid wastes derived from the large urban agglomerations. The sea is, unfortunately, becoming a repository for such materials on an increasing scale; while it is not easy to make a reasonable estimate of the amounts involved, Tables 3.3 and 3.4 indicate the magnitude of that part of the problem arising from wastes.

Table 3.2 Inorganic chemicals to be considered in water quality criteria for aquatic life in the marine environment

Elements	Equilibrium species (reaction)	Natural concentration in seawater[a] (μg/l)	Pollution categories[b]
Aluminium	$Al(OH)_3$, solubility of Al_2O_3 approx. 300 μg/l	10	IV c
Ammonia	NH_3, NH_4^+		IV c
Antimony	$Sb(OH)_6^-$	0·45	IV c?
Arsenic	As_2O_3 is oxidized to $HAsO_4^{2-}$	2·6	II c
Barium	Ba^{2+}	20	IV c
Beryllium	$Be(OH)_2$, solubility of BeO approx. 10 μg/l	0·0006	IV c?
Bismuth	$Bi(OH)_3$, solubility of Bi_2O_3 is unknown (low)	0·02	IV c?
Boron	$B(OH)_3$, $B(OH)_4^-$	$4·5 \times 10^3$	IV c
Bromine	Br^0, HBrO, Br^-	$6·7 \times 10^4$	IV c
Cadmium	$CdCl^+$, $CdCl_2$, $CdCl_3^-$ (the last two are probably the main forms)	0·02	II c
Calcium	Ca^{2+}	$4·1 \times 10^5$	IV c
Chlorine	Cl^0, HClO		IV c
Chromium	$Cr(OH)_3$, solubility of Cr_2O_3 unknown (low)	0·04	IV c?
Cobalt	Co^{2+}	0·4	IV c
Copper	Cu^{2+}, $CuOH^+$, $CuHCO_3^+$, $CuCO_3$ (probably main form) $CuCl^+$, complexed also by dissolved amino acids	1	IV c
Cyanide	HCN (90%), CN^- (10%)		III c
Fluoride	F^- (50%), MgF^+ (50%)	1340	IV c
Gold	$AuCl_2^-$	0·01–0·02	IV c
Hydrogen ion (acids)	$HCl + HCO_3^- \rightleftharpoons H_2O + Cl^- + CO_2$ $H_2SO_4 + 2HCO_3^- \rightleftharpoons 2H_2O + SO_4^{2-} + 2CO_2$	pH = 8 (alk = 0·0024 M)	III c
Iron	$Fe(OH)_3$, solubility of FeOOH approx. 5 μg/l	10	IV c
Lead	Pb^{2+}, $PbOH^+$, $PbHCO_3^+$, $PbCO_3$, $PbSO_4$, $PbCl^+$ (probably main form)	0·02	I a
Magnesium	Mg^{2+}	$1·3 \times 10^6$	IV c
Manganese	Mn^{2+}	2	IV c
Mercury	$HgCl_2$, $HgCl_3^-$, $HgCl_4^{2-}$ (main form)	0·1	I b
Molybdenum	MoO_4^{2-}	10	IV c
Nickel	Ni^{2+}	7	III c
Nitrate	NO_3^-	$6·7 \times 10^2$	III c
Phosphorus	Red phosphorus reacts slowly to phosphate $H_2PO_4^-$ and HPO_4^{2-}		
Selenium	SeO_4^{2-}	0·45	III c?
Silicon	$Si(OH)_4$, $SiO(OH)_3^-$	3×10^3	IV c
Silver	$AgCl_2^-$	0·3	III c

Table 3.2 Continued

Elements	Equilibrium species (reaction)	Natural concentration in seawater ($\mu g/l$)	Pollution categories
Sulphide	S^{2-}		II c
Thallium	Tl^+	0·1	III c
Titanium	$Ti(OH)_4$, solubility of TiO_2 unknown (low)	2	IV b?
Uranium	$UO_2(CO_3)_3^{4-}$	3	III c
Vanadium	VO_3OH^-	2	IV a?
Zinc	Zn^{2+}, $ZnOH^+$, $ZnCO_3$, $ZnCl^+$ (probably main form)	2	III c

[a] These values are approximate but are representative for low levels in unpolluted seawater.
[b] I–IV indicate order of decreasing toxicity; a—worldwide, b—regional, c—local (coastal, bays, estuaries, single dumpings).
? Indicates some question of the ranking as a menace and/or whether the pollutional effect is local, regional or worldwide.

Table 3.3 Ocean dumping by the United States: types and amounts, 1968.

Waste type	Amounts (tons)			
	Atlantic	Gulf	Pacific	Total
Dredge spoils	15,808,000	15,300,000	7,320,000	38,428,000
Industrial wastes	3,013,200	696,000	981,300	4,690,500
Sewage sludge	4,477,000	0	0	4,477,000
Construction and demolition debris	574,000	0	0	574,000
Solid waste	0	0	26,000	26,000
Explosives	15,200	0	0	15,200
Total	23,887,400	15,966,000	8,327,300	48,210,700

Table 3.4 Estimated polluted dredge spoils dumped off the coast of the United States.

	Total spoils (tons)	Estimated percent of total polluted spoils	Total polluted spoils (tons)
Atlantic Coast	15,808,000	45	7,120,000
Gulf Coast	15,300,000	31	4,740,000
Pacific Coast	7,320,000	19	1,390,000
Total	38,428,000	34	13,250,000

Despite the scale of these operations, there is surprisingly little information regarding their impact; most of this is ecological in character, although combined ecological and experimental studies would be more useful. It is proposed therefore to look briefly at the present position, and then to examine in some detail an instance from the Solway Firth.

Iron waste deposits

According to *Water Quality Criteria* it is the experience of American workers who have undertaken extensive studies of the disposal of acid-iron wastes in the New York Bight that, although this process has been in progress for two decades, no adverse effect upon the biota has been detected. Apparently the acid is rapidly neutralized by the seawater, and the iron is precipitated as ferric hydroxide. This ferric hydroxide is in the form of a flocculent precipitate which occurs in a measurable accumulation only close to the specified dumping ground. Indeed these so-called "acid grounds" are favoured by the local fishermen.

Sludge deposits

On the other hand, sewage sludges have a solid fraction which comprises about 5% of the total: this 5% contains some 55% of organic matter and 45% of aluminosilicates, but carries a trace-metal burden at least 10 times that of the natural sediments. Such sludge has been dumped in New York Bight since 1924 and, as a result, normal benthic communities have been eliminated from an area of about 10 square miles, and altered over an area of about 20 square miles. These sludge deposits have been shown by chemical analysis to carry a heavy burden of trace metals and petrochemicals. Moreover, it has been found that a fin-rot disease occurs in fish taken from the area, while the lobster *Homarus americanus* and the crab *Cancer irroratus* suffer from a necrosis of the exoskeleton induced by contact with this sludge deposit (which test animals tend to avoid when given the choice in the laboratory). In Britain, perhaps the best-studied area in which sewage sludge is dumped is that investigated by the Clyde River Purification Board off the Garroch Head in the Firth of Clyde. Here too, an increase in the content of organic carbon and of some trace metals is accompanied by a change in the benthic biota, and overall gross changes are limited to an area of about 20 km^2 of seabed; but little is known of the overall effects of toxic wastes, particularly at threshold levels.

Trace elements and Nereis

Some light has been thrown on this problem by work in the south-west of England, where Bryan and Hummerstone (1971; 1973a, b) examined the relationship between the occurrence of trace metals in the polychaete worm *Nereis diversicolor* and the sediment which it inhabits in estuaries subject to a variable degree of the influence of mining of non-ferrous metals (figure 3.1). It will be seen that the ratios of the concentrations of copper in the animals and sediments are similar; the two concentrations may be directly related. On the other hand, there was no obvious relationship between the concentrations of Zn, Pb, Mn and Fe in *Nereis* and those

INORGANIC WASTES

Estuary	Ratio—Body concentration: soil concentration				
	Cu	Zn	Pb	Mn	Fe
Plym	0·68	0·59	0·13	0·16	0·06
Dart	0·50	1·16	0·03	0·05	0·01
Avon	0·63	1·78	0·10	0·04	0·02
Camel	0·42	1·27	0·03	0·03	0·03
Tamar	0·24	0·32	0·02	0·03	0·01
Restronguet Creek	0·38	0·09	0·01	0·01	0·01

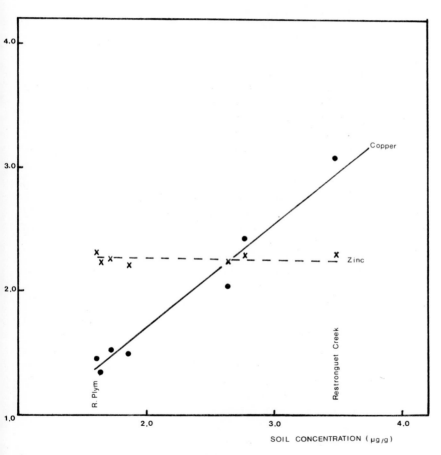

Figure 3.1 The relationships between the trace-metal content of *Nereis diversicolor* and the sediment which it inhabits in estuaries of south-west England.

found in the sediments; furthermore, the concentrations of zinc in the animal were remarkably constant, despite marked variations in the sediment concentration. This may be due either to differences in the concentration of each actually available to the worm, or the worm may regulate the concentrations of the other metals. The animals have developed a tolerance to both copper and zinc, a feature which is not readily gained or lost, and may indeed be a factor which is genetically controlled. There has been plenty of time for such a tolerance to have developed, since some of these estuaries, e.g. Restronguet Creek, have been receiving mining waste for at least 200 years. Whatever may be the truth regarding the other metals, it seems clear that this element is much more readily available in those sediments which have a high copper concentration than in those which do not.

China clay
Another industry which has contributed inorganic waste to the sea is the mining of china clay in Cornwall. This activity has gone on since the mid-eighteenth century, and the river-borne waste which resulted was carried to and deposited in the beds of Mevagissey and St. Austell Bays (figure 3.2). The input of waste reached 100,000 tons per annum in 1945 and increased to 245,000 tons per annum in 1962, after which a more rapid increase occurred, until a maximum level of 700,000 tons per annum was attained in 1967; this level of discharge was maintained until May 1970, when a reduction was brought about by the Cornwall River Authority. This reduction continued progressively until January 1973, when a new approach to the problem resulted in a total cessation of these discharges.

Surveys carried out in the period prior to final cessation of the discharge showed effects upon the biota of Mevagissey Bay in four zones. Close inshore, a narrow *azoic area* characterized by a severely impoverished fauna was found; here the most conspicuous members of the macrofauna were occasional specimens of the polychaete worm *Nephtys hombergii*. The remaining part of the area lying within a line drawn from Black Head to Penare Point, i.e. the *inshore area*, was characterized by the bivalves *Nucula turgida* and *Tellina fabula*. Here certain species which might have been expected to be present were notably absent (such as the polychaete worms *Melinna palmata* and *Pectinaria auricoma*, the bivalve mollusc *Phaxas pellucidus*, the gastropod mollusc tower shell *Turritella communis*, and the brittle starfish *Amphiura filiformis*). In June 1971 the area was characterized by a mass occurrence of the young of the sea cucumber *Labidoplax digitata*.

To seaward of, and based upon, the line drawn from Black Head to Penare Point, lay the roughly triangular *central area* which represented

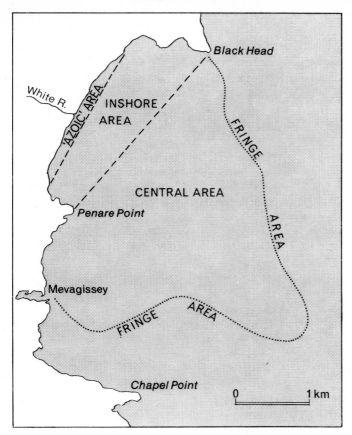

Figure 3.2 Faunal zones in Mevagissey Bay (after Probert, 1975).

the greater part of the deposited china-clay waste. This central area was characterized by an *Echinocardium filiformis* community (key species *Echinocardium cordatum*, *E. flavescens* and *Amphiura filiformis*) which commonly contained the sea anemone *Edwardsia callimorpha*, the bivalve molluscs *Thyasira flexuosa* and *Dosinia lupinus*, and the opisthobranch mollusc *Philine quadripartita*, in addition to those species noted above as conspicuous absentees from the inner area.

Finally, there was a *fringe area* in which the characteristic sediment and bottom fauna of the preceding region merged into the surrounding muddy sands and gravels. A more varied fauna was found here, but this community could be typified by the polychaete worm *Owenia fusiformis* and the sea cucumber *Cucumaria elongata*.

Probert (1975) considered that the factors contributing to the faunal impoverishment of the inner region of Mevagissey Bay were: the micaceous nature of the waste, and the interrelationship of the mica content, the sedimentation rate, and the shear strength of the sediment. These factors indicated that the rapid deposition of a micaceous sediment in Mevagissey Bay would be liable to result in a particularly incohesive unstable deposit in near-shore regions.

The Solway Firth

The writer and co-workers have been engaged on studies of the Solway Firth, particularly in relation to waste disposal, since 1961. It is believed that this work can now contribute something to the understanding of estuaries in general, as well as to the cases discussed above.

The Solway Firth is a large coastal-plain estuary, tributary to the north-east Irish Sea. It is of particular interest in that it is subject to little direct influence from industrial activity, except in that portion of coastline which lies between St. Bees Head and Maryport (see figure 3.3). Nevertheless, this relatively short length of coast was one of the principal cradles of the Industrial Revolution, and in particular of the iron and steel industry. Indeed, its importance was derived not so much from the necessary proximity of coal, limestone and iron ore, but from the fact that the iron ore, haematite, was of a quality particularly suitable for use in the Bessemer process which was developed here. This part of Cumbria, which was to do so much for the rest of the world, is only now emerging from the blight of having been so early on the industrial scene and therefore having worn out its natural resources. This re-emergence is responsible for the present studies and, paradoxically, it is these studies which are examining the changes in the marine environment brought about by our vigorous but unheedful forebears.

Slag disposal

In the production of iron and steel, vast amounts of slag (consisting largely of calcium silicate) are produced as a waste product (Table 3.5). In present-day practice, these slags are cooled before being deposited upon spoil heaps or bings; in former times, however, the slag was poured onto the bings while still hot. On the coast of Cumberland, the early coal and iron masters were evidently intolerant of spoil dumping on land which could be put to another use, when the convenient seashore was nearby. Those bings derived from coal mining had a sulphur content which was large enough to undergo spontaneous combustion. On the other hand, the hot slag solidified on contact with the shore and formed an artificial rock known

INORGANIC WASTES 83

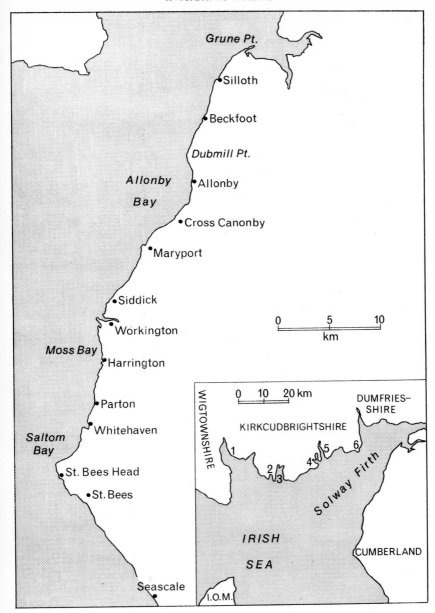

Figure 3.3 Map of the Cumberland coast. Inset of Solway Firth area showing stations on the north shore thus: 1—Carsluith; 2—Brighouse Bay; 3—Kirkcudbright Bay; 4—Auchencairn Bay; 5—Urr Water and Rough Firth; 6—Southerness Point.

Table 3.5 The composition of slagcrete and old coal-bing samples taken from the shore at Siddick and Harrington, 1975, compared with some components of normal beach sands recorded in the literature. N.B. Selective, not total analysis.

Major components	Composition (%)		Beach sands
	Slagcrete/coal bing		
	Mean	Range	
Si	45·1*	11·2 –78·5	87·87 –99·72**
Fe_2O_3	17·70	1·6 –72·2	0·7000– 1·20
CaO	20·43	2·66–39·88	0·0500– 1·30
Al_2O_3	5·7	0·3 – 8·6	0·0027– 6·60
MgO	3·81	1·33– 5·47	0·0200– 0·54
Mn	1·28	0·3 – 3·63	—
SO_4	0·80	0·18– 1·63	—
Ba	0·53	0·03– 2·51	—
$S^=$	0·41	ND– 1·02	—
PO_4	0·12	0·04– 0·27	—
Trace metals		Composition (ppm)	
	Mean	Range	
Cd	6·63	3–10	—
Cr	108	50–250	—
Cu	74	15–200	—
Pb	544	100–2500	—
V	125	70–200	—
Zn	304	30–900	—

* Si as SiO_3 ** Si as SiO_2

to the industry as *slagcrete*. Traces of old coal bings containing the characteristic fragments of shale burned by spontaneous combustion may be found as far down the shore as the mean tide level (MTL), and there is reason to believe that slagcrete was sometimes poured as far down as the low water mark (LWM). Both obliterated the resident shore life.

Such mass destruction is beyond dispute, as indeed is the destruction still being wrought by the National Coal Board who insist on using the seashore as a dustbin; the abrasion which results from transport of this waste by longshore drift scours every living organism from all the rock surfaces in its path. Indeed, such damage is so patent that, apart from the obvious comment that we ought to be better informed and better behaved than our forefathers, there is nothing here of scientific interest to discuss.

What is of great interest, however, is the fact that only remnants of these early spoil heaps can be found on the shore today and, even as late as the mid-1960s, one of the two bings above the high water mark at Siddick was bulldozed into the sea. Coarse material from both types of bing, including shale, burnt shale, slagcrete, refractory brick and lumps of ferruginous material which were clearly derived from a blast furnace, are found in the high shore level "shingles" as far upstream as Grune Point, 20 miles from Maryport, the site of the nearest blast furnaces, from which it was carried by longshore drift. Even at Beckfoot, 12 miles north of Maryport, samples of "shingle" taken from the swash zone (the interface between land and shore) may contain 7% of such material, and a conservative estimate would suggest that this zone north from Maryport to Grune Point contains about 2×10^6 metric tons of it.

Chemical content

This conclusion applies only to the particles which are of a size greater than 5 mm diameter, but all the finer sediments contain readily recognizable particles of all sizes from 5 mm to 0·0625 mm diameter (i.e. the upper limit of the silt-clay fraction), both on the shore and for not less than 4 miles below the low water mark in Allonby Bay (figure 3.4). While it is fair to say that the analyses quoted with respect to beach sands in Table 3.5 represent rather too simple a picture, it will be noted that the iron content of the slags is very different from that of normal shore sediments, and is accompanied by high concentrations of manganese (1·28%, range 0·3–3·63%) and of the trace metals cadmium, chromium, copper, lead, vanadium and zinc—all of which are known either to be vital to living systems or are biocidal. Of these metals, analyses for iron, copper, manganese, lead and zinc on the shores at Parton, Siddick, Cross Canonby, Allonby and Beckfoot are presented in Table 3.6. Of these shores, Parton and Siddick may be regarded as sites at which a considerable direct input has occurred or which have received material from adjacent sources —so much so that the "sands" at Parton are black from the presence of coal and shale, and those at Siddick are a rusty brown colour.

The scarcity of the lugworm

One of the interesting features of a series of shore transects (carried out annually since 1967) is the relative paucity of the lugworm *Arenicola marina* at Siddick (Table 3.7). In the years 1967–1970, the populations were so sparse that a density of 0 per m^2 was recorded on the transect, although it was probably of the order of ≤ 1 per $20\,m^2$ for the shore as a whole. This pattern of very low maximum density began to change in 1971, though it tended to remain low until 1975. In 1972, it was possible to show that although the density of *Arenicola* was low, it was apparently related to the iron and trace-metal content of the soil, since no *Arenicola* were present (and indeed never occurred) in the inter-reef sediment, which always has the highest content of iron and trace metals on this shore. The maximum density of 2·5 worms/m^2 occurred at low shore levels, where

Table 3.6 The concentration of certain trace metals in the shore sediments at Pa

Shore	Trace metal			
	Fe (%)		Cu (ppm)	
	Mean	Range	Mean	Range
Parton	5·02	2·7 –7·3	91	55–170
Siddick (Interreef)	4·7	2·8 –6·5	52	29–75
Flats	2·2	1·5 –3·0	23	15–29
Cross Canonby	2·3	1·7 –3·6	32	19–60
Allonby	2·2	1·75–4·9	25	16–53
Beckfoot *Arenicola* flat	1·2	0·95–1·55	11	5–18
Ardmore	1·0	0·7 –1·2	9·5	6–15

the iron and trace-metal content was lowest (figure 3.4). If these data are compared with those for Parton and the stations upstream which receive this material indirectly, a good ecological case can be made for suggesting that these materials of industrial origin have an adverse effect upon the lugworm *Arenicola*. This case is strengthened when it is appreciated that, whereas the soils at Parton and at the inter-reef area at Siddick contain no *Arenicola*, the sediments of a similar-grade structure at Brighouse Bay on the coast of Kirkcudbright are inhabited by the lugworm; and strengthened even more when we examine more closely the known effect of iron upon this animal. It was shown by Reid in 1930 that the lugworm is repelled by, and will not burrow in, soils containing $2·0\%$ Fe_2O_3 in an amorphous state. When this oxide is suspended in seawater at a concentration of $0·21\%$, it forms with the body mucus an envelope which prevents

Table 3.7 Maximum density of the lugworm *Arenicola marina* recorded in the annual survey of Cumberland shores, 1967–75. (Survey normally in September; exception August 1973.)

Year	Parton middle shore	Siddick	Cross Canonby	Allonby	Beckfoot
1967	0	0	16 (28)	12	12
1968	0	0	24 (64)	16	24
1969	0	0	12	20	24
1970	0	0	4 (16)	16	20
1971	0	2·8	10	7·2	10
1972	0	3	10	12	10
1973	0	7·6	9·6	7·6	13·8
1974	0	4	12·4	7·2	16·7
1975	0	10	9·2	*	e 20**

*Count impossible due to wind-blown water sheet.
**Casts affected by wind; approximate estimate only.

INORGANIC WASTES

dick, Cross Canonby, Allonby and Beckfoot, and compared with Ardmore, Firth of Clyde.

	Trace metal				
Mn (ppm)		Pb (ppm)		Zn (ppm)	
Mean	Range	Mean	Range	Mean	Range
1312	600–2300	73	55–100	110	80–155
6809	4000–9600	133	100–160	290	200–380
1430	1200–2400	47	40–70	104	80–135
2445	1600–5800	62	30–95	113	90–175
2430	1650–4700	63	50–100	114	92–200
1162	450–1900	41	25–60	76	33–150
187	70–800	37	20–55	50	30–65

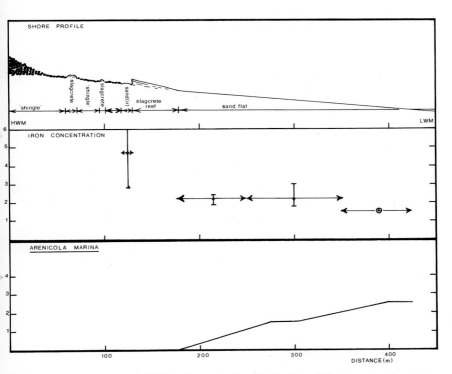

Figure 3.4 Transection of Siddick shore showing distribution of the lugworm *Arenicola marina* in relation to the concentration of iron in the shore sediments. Sand (ir) = sand in inter-reef region.

contact with the surrounding aerated water, and thus kills the animal. It is noteworthy that many of the soils examined, including those from Parton and Siddick, are liable to deposit Fe_2O_3 on the walls of the sample jar, and that the supernatant liquor samples from Parton, preserved with formalin, may become heavily burdened with Fe_2O_3. These observations suggest that the iron is probably loosely bound if not freely available. We may therefore have good grounds for believing that these apparently inert wastes have an adverse influence upon the lugworm, and that the introduction of excessive amounts of iron may be responsible for the observed differences in distribution. Unfortunately, this attractive and deceptively simple hypothesis can be disproved.

Disproof comes from three sources—one ecological and two experimental in nature. The ecological consideration derives from the number of species of macrofauna taken on the standard transect in the period 1967–74. After the first two years, the number of species present declined from 9 or 10 to 3; but in 1972 a marked increase became evident, and by 1974 the greatest number of species, i.e. 17, was recorded; moreover, the species composition was the same as that at stations upstream. This increase in the number of species was not accompanied by a decline in the iron and trace-metal concentrations, which tended to increase from 1972 to 1974. It could have resulted either from the working of long-term cycles of abundance or (as seems more likely) because a shingle bank which had been building up around an effluent line by now extended to the north and sheltered these sands from wave action, thus allowing the greater development of infauna. Though the maximum abundance of the lugworm tended to remain low up to and including 1974, by 1975 it was evident that while in the very short term its abundance may have been highly variable, it could, given the appropriate conditions, attain a considerable density in the shore at Siddick.

The experimental evidence was derived from two sources—of which the first was not initially related to this problem. The effluent line noted above released waste which contained lignified and non-lignified wood pulp fibres. While the dispersal from this plant was such that no environmental problems were evident, it was important to have some indication of the rate at which such materials are likely to be assimilated in the environment. This screening of cellulolytic activity was carried out by measuring the time taken to break a number 36 gauge cotton thread exposed to soil micro-organisms in the manner shown in figure 3.5. In a general screening of the soils from the Solway Firth and the bed of the N.E. Irish Sea, it was found that cellulolytic activity was greatest in the more silty soils, either from the bed of the Irish Sea or from the accretion zones of salt marshes; that the activity of the less silty soils was a function of the amount of water present, and that those from Siddick were generally similar in activity to those of a similar-grade composition from elsewhere. Furthermore, and unlike the sulphate-reducing bacteria, the organisms respon-

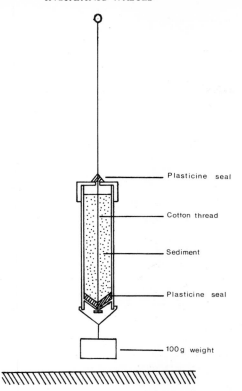

Figure 3.5 Method of exposing a cotton thread (No. 36 gauge) to a sediment as a measure of cellulolytic activity. Total weight suspended = 150 g.

sible appeared to be rather sensitive to environmental change. In order to test for a possible poisoning of the cellulolytic micro-organisms by the material derived from the bings of Siddick, aliquots of soil from the inter-reef area, which have a characteristically high iron and trace-metal content, were mixed with soil from the active accretion zone of Auchencairn Bay, Kirkcudbrightshire, in proportions up to 80% Siddick : 20% Auchencairn Bay. For comparison, similar mixtures were prepared using soil from Brighouse Bay, Kirkcudbrightshire, which is similar in grade composition but, like Auchencairn Bay, is far removed from the influence of bing-derived materials. The results obtained are shown in figure 3.6. It can be seen that no adverse effect upon the cellulolytic activity can be ascribed to the bing-derived material from Siddick.

The next phase in the investigation determined whether the lugworm *Arenicola* is overtly influenced either directly by bing material or by sedi-

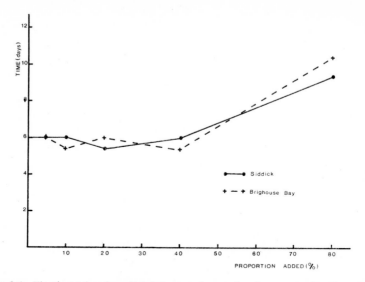

Figure 3.6 The time taken for cellulolytic organisms to break a cotton thread exposed to soil from Auchencairn Bay, diluted with varying amounts of soil from Siddick (inter-reef area) and Brighouse Bay. Incubation at 20°C.

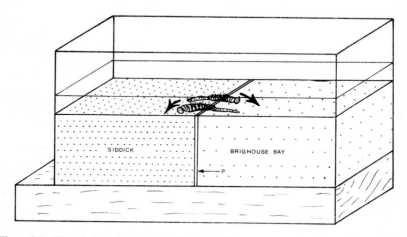

Figure 3.7 The choice chamber in which the lugworm *Arenicola marina* was offered sediment either derived from bings or from Brighouse Bay.
Perspex boundary. Arrows indicate the direction followed when placed across boundary with head upon one substratum and tail upon the other.

ments from which it is normally absent in the shores at Parton and Siddick. These materials alone were first offered as a prospective habitat to lugworm taken from sediments at Ardmore, Firth of Clyde, where they are abundant (51 per m^2). They have a relatively low iron and trace-metal content, thus excluding any question of adaptation to high trace-metal concentrations. Such worms showed no aversion to these test sediments, and lived in them for 8 weeks without mortality. Furthermore, similar results were obtained with the polychaete worm *Nephtys*, the bivalve molluscs *Cerastoderma*, cockle and *Macoma balthica*, the shrimp *Crangon crangon*, and the shore crab *Carcinus maenas*, all of which habitually use sediment as a shelter. The sand goby *Pomatoschistus minutus*, the hermit crab *Eupagurus bernhardus*, the sea urchin *Psammechinus miliaris*, and the starfish *Asterias rubens* lived above such sediments without apparent inconvenience.

Having shown that there is no obvious repulsion demonstrated by the lugworm and other organisms towards these soils, the next step was to offer the worm a choice of substratum. By partially filling a tank with test and control soils, a worm could be laid across the boundary with either its head upon a bing-derived sediment or upon that from Brighouse Bay (figure 3.7). Seven pairs of worms were tested in each of the Siddick Inter-reef/Brighouse Bay and Parton/Brighouse Bay combinations: in no case was rejection demonstrated by animals placed with their heads upon either of these bing-derived sands, but two placed with their heads upon Brighouse Bay sand rejected it in favour of burrowing into inter-reef material. While the animals subsequently moved freely across the boundary and, on balance, apparently preferred inter-reef and Parton to Brighouse soils, no such relationship could be demonstrated statistically.

Similar tests using the cockle *Cerastoderma*, the starfish *Asterias rubens*, and the hermit crab *Eupagurus bernhardus* failed to reveal any selection against inter-reef soil and freshly eroded bing material from Siddick, and shore material taken from Parton. We must therefore conclude that the original hypothesis (based solely upon ecological data)—that materials which were eroded and eroding from coal and blast-furnace bings could have an adverse effect upon the fauna of sandy shores when reduced to particles of a comparable size to that which is normally inhabited—has no foundation in fact. This finding indicates that conclusions derived from short-term ecological studies must be treated with caution, and should where possible be given experimental support. Moreover it is evident that measurements of total trace-metal content have no value in relation to living systems; rather it is the measurement of *available trace-metal concentration* which should be given a high priority in this field. There is one further corollary to this conclusion—all dumping is not bad. This is an interesting conclusion since, in many ways justly, dumping has been very much under attack in recent years. Truly inert materials may be safely introduced into the marine environment, once they have been cleared by screening methods which are either available or can be readily devised; but the malpractices of mining, still evident on the coasts of Durham, Cumbria and elsewhere, do not fall into this category.

However, the problem remains that the populations of lugworm at Siddick have been variable in abundance, have rarely matched those of the upstream stations, and that the shore at Parton never has any lugworm present. In the latter case (and in the earlier years at Siddick, i.e. before the development of the shingle bank to its present state) the only reasonable conclusion left is that wave action creates conditions which are unfavourable for the development of abundant populations of lugworm. In more recent times it would appear that the effect of intensive bait-digging cannot be ignored at Siddick. Strong support comes for this view from the last survey carried out in 1975, when participation in sea-fishing competitions elsewhere had freed this shore, for about two weeks, from the attentions of the bait diggers who frequently reduced it to the appearance of a battlefield. This period was long enough for the immigration of individuals from outside populations and enabled us to record the greatest abundance of *Arenicola* here in 9 years of work.

Phosphogypsum

Calcium sulphate is a by-product of many industrial activities, of which the most important is the wet-process method of phosphoric acid production, where every ton of P_2O_5 produced is accompanied by about 5 tons of gypsum (phosphogypsum). While a number of outlets for this gypsum are possible, the availability of a purer dry form (from gypsum mines) means that these outlets are not as forthcoming as a superficial consideration of the problem would indicate. Since phosphoric acid is of major importance in the production of fertilizers (about 65% of all the phosphoric acid produced in the United States is used in this way) and widely used throughout the chemical industry for the manufacture of products ranging from cleaning agents to food additives and pharmaceuticals, it is produced in large quantities. In European countries, a large part of the production is by the wet process and it follows that the gypsum waste product is significant in quantity if not in effect. Thus, the practices used to deal with gypsum range from the construction of large spoil heaps in countries with a low rainfall to release into the sea in Britain and other European countries. If we consider the British case specifically then, even if the large white spoil heaps were an acceptable addition to the landscape, our high rainfall would dissolve much of the deposited material, and a permanent increase of hardness due to the calcium sulphate would be experienced by the adjacent surface waters. An understanding of the nature of the impact of this substance upon the quality and the biota of the receiving waters is therefore important.

For this reason, an investigation has been in progress since 1970. For the past 20 years, a factory at Whitehaven (figure 3.3) has produced about 1500 tons per day of phosphogypsum, and this, together with other effluent materials, was introduced into the sea at the high water mark (although in the light of this survey the practice has been modified markedly). Apart from the liquid effluent itself, the most obvious manifestation of its presence was the development, on the shore, of a small white delta, of a size which varied with the weather conditions, but which had a length of 25 m north and south of the inflow at the low water mark and which narrowed across the shore (which has a width of 60 m at this point); there was some small extension of this delta into the sublittoral zone. Bearing in mind the duration of this practice and the amount of gypsum involved, we might consider that the sediments at the seabed would undergo a marked and inevitable accumulation of $CaSO_4$.

Survey results

However, the results obtained are typified by those in Table 3.8. Starting in Saltom Bay, which receives the effluent, it was found that there was no obvious gradient of sulphate concentration as we move further away from the delta; the change from delta to normal seabed sediments was sharp, and the sulphate concentration in the sediments was both relatively even and much lower than that of the delta—0·11% as compared with about 50%.

Secondly, the concentrations of sulphate increased only at distances more than 1 mile from the coastline, and in a different type of sediment lying at a greater depth and outside the immediate influence of the outfall. At this point it became clear that any norm with which these results could be compared was lacking, and an extended survey of both littoral and infralittoral sediments was necessary (see Table 3.8). Even in the sediments at the low water mark in Saltom Bay, and at a distance 200–600 m south of the outfall, the sulphate concentration was remarkably low compared with these other situations. Although transport of undissolved calcium sulphate away from this source remained a possibility, nevertheless we were forced to agree with the contention which had long been made that, with a solubility of 0·4% in seawater as compared with 0·2% in fresh water, the calcium sulphate was being brought into contact with a very considerable volume of seawater, was dissolving, and was by that

Table 3.8 The concentration of sulphate in the sediments of the Solway Firth area.

		Sulphate concentration (%)	
Source	n	Mean	Range
Delta below outfall		ca. 50%	
Saltom Bay			
Shore 200–600 m south of outfall	8	0·11	0·08–0·14
sublittoral			
(i) Area 2 miles long × 1 mile offshore off outfall	23	0·11	0·07–0·26
(ii) Area 5 miles long × 2 miles offshore off outfall	90	0·19	0·06–0·53
Allonby Bay			
sublittoral	26	0·11	0·07–0·24
Kirkcudbright shore			
Auchencairn Bay (a) Seaside	16	0·16	0·12–0·24
(b) Balcary Bay	13	0·23	0·13–0·32

n = number of samples.

means removed from the seabed. Indeed we can look at this another way: if we assume total solution and a conservative tidal excursion of 3 km, a mean depth of 10 m, and a contribution to a body of water 100 m wide, then the maximum concentration of sulphate likely to result is 0·02% per tidal cycle.

Concentration of sulphate

The next step in the investigation was to examine the concentration of sulphate in the seawater which received the effluent. Here, however, we cannot compare the concentrations directly, because the salt content of seawater is variable and depends upon the amount of fresh water run-out from the land, evaporation and the degree of mixing with waters of a more oceanic origin. What can be done is to compare a known parameter of seawater with the sulphate concentration in each sample. First, however, let us consider the known composition of seawater.

The amount of dissolved inorganic matter, expressed as grams per kilogram of seawater, is known as the *salinity* (S). In the open ocean, this value approximates to 35 g/kg, and it was one of the dogmas of oceanography that the major ions in seawater are present in constant proportions. Indeed, it was upon this basis that the early determinations of salinity and, in effect, all subsequent ones, were made. It is relatively easy to measure the amount of chloride plus the chloride equivalent of bromide and iodide present by means of a titration using silver nitrate. The total weight of these elements per kilogram of seawater is known as the *chlorinity* (Cl). To derive the salinity from the chlorinity, earlier workers used the Knudsen formula:

$$S = 0\cdot030 + 1\cdot8050\,Cl\,(g/kg)$$

although this did introduce an error due to the salinity of the river waters flowing into the Baltic Sea (i.e. the term 0·030) upon which the formula was based. A Joint Panel of Experts appointed by UNESCO later amended the formula to

$$S = 1\cdot80655\,Cl\,(g/kg)$$

In an important review of this subject (Wallace, 1974) it was noted that no one has ever demonstrated that these major constituents exist in constant proportions, and even Knudsen did not consider his definition to be anything other than a provisional one.

Interpretation of data

Because of the variations in salinity, the data on sulphate in relation to the chlorinity were interpreted by adopting the following procedure. If we assume first that the sulphate and the chloride (plus bromide and iodide) ions occur in a constant proportion to one another, and if we assume that any change in salinity is due to a dilution with distilled water, then the ratio of sulphate to chloride concentration expressed as percentages of normal seawater ought always to be unity. In more practical terms, the rocks over which fresh waters flow into the Solway Firth contain significant amounts of gypsum (indeed gypsum was once quarried from the cliffs of Saltom Bay, and subterranean mines are still present in this area). With increased dilution by fresh water, the ratio ought to depart from unity in favour of a greater proportion of sulphate present. However, in the Outer Solway and N.E. Irish Sea area, where the salinity is always high, i.e. $\geqslant 30\,g/kg$, this ratio is 0·94 (range 0·86–1·06) and rather less sulphate is present than we would expect; likewise in Loch Fyne, where high salinities always prevail, this ratio is 0·92 (range 0·89–0·96). In Allonby Bay, the station farthest upstream in the Solway, both the lowest and the highest values of any of these stations were recorded and ranged from 0·78 to 1·48, average 1·00. We must therefore conclude that, even in situations far removed from an input of calcium sulphate, such as that under discussion, there is no good evidence for a constant proportionality of sulphate with respect to chlorinity.

Once again it has become apparent that the investigator in this field is faced with the problem of establishing just what is the norm against which he can measure effects due to a source of waste. Whatever other preoccupations there may be, all investigations must devote a considerable effort towards establishing the base-line, since in most cases this is either unknown or very imperfectly known. Such long-term work must necessarily cover a wider area than that of the immediate problem but, without it, all the other projects must be regarded with reserve. In the case of Saltom Bay, the results quoted have been taken from a wider context which confirmed the conclusion that sulphate and chlorinity in seawater cannot be regarded as existing in a constant relationship. Nevertheless it must be concluded that, within a distance of 600 m north and south of the outfall, where the influence of tidal and longshore currents will be most marked, the sulphate content of the seawater at the tide edge falls within the limits found at situations far removed from this source. In the circumstances, therefore, it is not too surprising that the sublittoral fauna of Saltom Bay is typical of the same type of sediments elsewhere and includes plumose sea anemone *Metridium senile*, sand mason *Lanice conchilega*, acorn barnacle *Balanus crenatus*, shrimp *Crangon crangon*, masked crab *Corystes cassivelaunus*, swimming crab *Portunus holsatus*, hermit crab *Eupagurus bernhardus*, common starfish *Asterias rubens*, starfish *Astropecten irregularis* and brittle star *Ophiura texturata*. The manner in which mullet haunt and feed upon the rocks of the shore has already been noted (p. 75), and in 1975 the commercial fishery for sole *Solea solea* was particularly good in Saltom Bay. It must be recognized that all except the first three species listed are capable of active movement, and would therefore be expected to shun adverse conditions; likewise all except the mullet are closely associated with the seabed.

In this particular case, then, we have a situation in which it is difficult to demonstrate any effect either upon the infra-littoral substratum and the main body of the seawater or upon their inhabitants. Indeed, effects due to this effluent are apparently confined to the shore for a distance 600 m to the south of the outfall and to a less certain distance to the north, for it is here that the gross effects which result from the transport of National Coal Board shale by longshore drift are most evident. In addition, the water interface between land and sea, i.e. a band about 50 m wide from the tide's edge, is influenced by the development of longshore currents. Within this area of action, i.e. from high water mark to 50 m out from the low water mark, field observation and toxicity studies suggest that it is the pH which is of major importance.

Eutrophication

Eutrophication has become a by-word which is often used in a variety of connotations unrelated to the meaning ascribed to it by scientists. The simplest definition of eutrophication is the process by which enrichment with nutrients occurs. In this process, nitrogen and phosphorus are considered to be of primary importance—even though this is a gross oversimplification of the problem. Truly enhanced production is impossible without these nutrients, but other factors control both the quality of the growth which is produced and the nature of its impact upon aquatic biota and man himself. Possibly this is one of the more important attributes of inorganic waste disposal. There is no doubt, for example, that inorganic mercury compounds are in themselves inimical to many living organisms, but it was the transformation of such inorganic compounds, by micro-organisms, into the much more toxic and available methyl mercury that brought disaster upon the fishing community at Minimata.

The popular response to the whole matter of eutrophication has been influenced by the untoward effects in fresh water where (particularly when sewage is involved) massive blooms of blue-green algae have brought about fish death by depletion of oxygen, particularly when the excessive blooms decay. Even while still healthy, the blooms release toxic substances to the water which kill fish and mammal alike, and at lower concentrations provoke the development of allergies in those human beings whose water supply is tainted with the toxin. In many ways, especially in North America, these events have led to the creation of an atmosphere of the witch-hunt, in which phosphorus as a nutrient has been cast as the devil, and in which the qualifications noted above have largely been ignored. To some extent this implication of phosphorus has spread to the marine environment, ignoring the fact that while it may be true that the supply of phosphorus is limiting in fresh water, it is nitrogen which is limiting in the sea.

"*Red tides*"

In general, the more unpleasant symptoms of eutrophication in the sea are the development of "red tides" in open waters, and extensive growths of the green algae *Ulva* and *Enteromorpha* upon the sand and mud flats may occur in estuaries which receive sewage wastes. However, the introduction of large amounts of nutrient do not necessarily give rise to either of these conditions or, if large filamentous algal masses do develop, they are not necessarily accompanied by the unpleasant symptoms noted at Ardmore (below). As long ago as the 1930s it was recognized that an abundance of nitrates and phosphates derived from London's sewage was present off the Thames estuary, and that an abundant phytoplankton resulted. Later, in the 1950s, it was shown that about 2900 tons of phosphorus per annum passed into the wedge of 171 cubic miles of seawater which lay against the east coast of England north to the Humber and gave rise to an increase of 4 mg of phosphorus per m^3. In this area, the catch per unit area is about double the corresponding catch in the rest of the North Sea, in the English Channel and in the Kattegat/Skagerrak region, and is about 25 times the catch for the Baltic Sea as a whole. Some two thirds of this increased catch is ascribed to the nutrients derived from London.

In contrast, in recent years there have been a number of outbreaks of red tide off the north-eastern coasts of the British Isles. These have been associated with a period of increased run-out and thus with an increase in the supply of nutrients. The red tide, which is known to occur in many parts of the world, arises when conditions are favourable to the development of a near monoculture of single-celled dinoflagellates such as *Peridinium* and *Gymnodinium*. These are organisms which are adapted to live in warmer waters having a low concentration of nutrients, but which can take advantage of an excess of nutrients in warmer waters to produce the red tide. The causative organisms secrete a neurotoxin which is very poisonous to man and animals alike, and may bring about a mass mortality of fish.

Green algae
It has equally long been known that, given an excessive supply of sewage nutrients, a massive growth of green algae can occur, especially in the littoral zone, where the tide eventually accumulates decomposing heaps, giving rise to the offensive smell of hydrogen sulphide. This occurred annually on the shores of the Firth of Clyde from Helensburgh to Cardross. The phenomenon was studied at Ardmore where, in the long hot still days of summer, the growth of *Enteromorpha* could flourish to such an extent that 13 acres were covered with large rotting mats of the alga in 1971, and a further area of 20 acres, which had only occasional thin mats of alga, was anaerobic to the extent that over 25–50% of it (5–10 acres) the black reduced sulphide layer had reached the surface and deposits of sulphur had occurred widely. In 1972, a more detailed investigation showed that, compared to the salt marshes and sand flats of the Solway Firth area which never demonstrated any of these unpleasant symptoms, the sands at Ardmore, which lack adjacent salt marshes, were deficient in both phosphate (0·11% compared with 0·20% in the Solway) and nitrate (0·14% compared with 0·44% in the Solway). The other outstanding difference between the two areas is the relative lack of sewage input to the north shores of the Solway. It should be noted that use of the sedimentary nutrient concentrations is proper here, since these reflect the nutrient status of the area as a whole, and that the phytoplankton bloom in coastal plain estuaries is controlled by the release of phosphate from the seabed.

Sewage treatment
Encouraged by an increasing public consciousness that sewage effluents should be properly treated before release to the sea, the Clyde River Purification Board designed and installed a treatment plant for the sewage

of Helensburgh. This plant releases only treated liquid effluent. 1975, the first year in which the plant worked, brought a severe test of its efficiency, for it was an unusually fine summer, accompanied by prolonged periods of high temperatures and lack of wind. Admittedly sewage plants do not lack odour, but the shore at Ardmore showed none of the excessive growth of *Enteromorpha*, extensive anaerobiosis or the stench of hydrogen sulphide so typical of earlier years. What then was different? The treated effluent still supplied both nitrate and phosphate, but the solid fraction of sewage was absent. The unpleasant aspect of eutrophication in estuarine waters is not due to the supply of phosphates or nitrates, as indeed the results from the Solway indicated, but to some factor associated with some other part of the sewage.

A hypothesis

We are now in a position to consider a hypothesis regarding the development of adverse algal blooms in coastal waters. The first point to recognize is that Ardmore has no associated development of salt marshes. The estuaries and coastal waters north of the Wash are similarly poorly endowed. The area from the coast of Kent northwards to the Wash, however, is (like the Solway Firth) well endowed with salt marshes; but, while the north shore of the Solway does not have a sewage input equal to that from Helensburgh, the phosphate and nitrate levels are higher in the sediments of these shores than at Ardmore. Considering the Thames estuary and the Solway together, it looks as if the salt marshes produce some substance or substances which promote an enhancement of biological productivity without the development of unpleasant symptoms (e.g. red tide or H_2S) in the presence of nutrient enrichment, be it due to sewage or to run-off from the land. In recent years it has been shown that the higher plants which live in salt marshes, e.g. cordgrass *Spartina* and eelgrass *Zostera*, absorb mineral phosphorus from the shore soil and transform it into organic compounds, which are then pumped out into the surrounding water as external metabolites. Similarly, the blue-green algae which also live in salt marshes can fix atmospheric nitrogen and liberate it to the seawater as amino acids; both kinds of compounds are important to the maintenance of high-quality water and healthy stocks of the biota in coastal waters. Indeed, it has long been realized that the organic materials present in seawater might be of great significance; but it remained for Lucas (1938) to postulate the Theory of Non-Predatory Relationships, which stated that:

In its lifetime an organism (plant or animal) secretes organic materials which might be either of advantage or disadvantage to associated organisms.

Since that first expression of the significance of external metabolites—which may be hormones, antibiotics or vitamins—the importance of these substances to life in the sea has become generally recognized.

With these points in mind, it is possible to construct a hypothesis.

In situations of high inputs of nutrient materials—particularly (but not exclusively) those derived from sewage—the salt marshes, by their ability both to use and to transform such materials, act as a safety valve to prevent the development of the unpleasant consequences of red tide in the main water body or excessive green algal growth on sand and mud flats. Where such salt marshes are absent, certain nutrient components associated especially with untreated sewage can, either in the absence of external metabolites derived from salt marshes or in the lack of the mechanism inherent in salt marshes to utilize these materials, promote the development of these symptoms; but in neither case is the ambient concentration of phosphate and nitrate the causative agent.

If this hypothesis is correct, then it would suggest that land-fill enthusiasts are antisocial in two ways.

(a) They affect amenity by laying the way open to the production of H_2S.
(b) They create a health hazard, which arises from the red tide, both to those with a predisposition to bronchitic complaints and to those liable to eat neurotoxin-contaminated fish.

In the light of this argument, it appears fortunate that the third London Airport at Maplin has been rejected.

Commentary

The examples which have been discussed were chosen partly because they constitute some of the research interest of the University of Strathclyde Marine Laboratory. They were also chosen because they demonstrate the essential subtleties of the systems with which biologists working in the field have to contend. This is a field in which there are no shortcuts and in which no simple analysis can give a correct solution to the problem. The investigator needs patience and endurance. The sponsor also needs patience, for it is his money, both in plant operation and in commissioning a study, which is at risk; but an investigation is of greater value only when it is carried through to the end.

FURTHER READING

Barrett, J. and Yonge, C. M. (1970), *Collins' Pocket Guide to the Sea Shore*, Collins, London. Most of the plants and animals referred to in the text are included; where a particular species is not represented, a type from its genus may be present.

Bryan, G. W. and Hummerstone, L. G. (1971), "Adaptation of the Polychaete *Nereis diversicolor* to Estuarine Sediments containing High Concentrations of Heavy Metals. I. General Observations and Adaptation to Copper," *J. mar. biol. Ass. U.K.*, 51(4), 845–863.

Bryan, G. W. and Hummerstone, L. G. (1973a), "Adaptation of the Polychaete *Nereis diversicolor* to Estuarine Sediments containing High Concentrations of Zinc and Cadmium," *ibid.*, 53(4), 839–857.

Bryan, G. W. and Hummerstone, L. G. (1973b), "Adaptation of the Polychaete *Nereis diversicolor* to Manganese in Estuarine Sediments," *ibid.*, 859–871.

Howard, T. E. and Walden, C. C. (1965), "Pollution and Toxicity Characteristics of Kraft Pulp Mill Effluents," *Tappi*, 48(3), 136–141.

Lucas, C. E. (1938), *J. Cons. Int. Explor. Mer.*, 13, 309.

National Academy of Sciences (1972), *Water Quality Criteria* 1972, Environmental Protection Agency, Washington, D.C.

Perkins, E. J. (1974), *The Biology of Estuaries and Coastal Waters*, Academic Press, London and New York.

Probert, P. K. (1975), "The Bottom Fauna of China Clay Waste Deposits in Mevagissey Bay," *J. mar. biol. Ass. U.K.*, 55(1), 19–33.

Wallace, W. J. (1974), *The Development of the Chlorinity/Salinity Concept in Oceanography*, Elsevier Oceanography Series, 7.

CHAPTER FOUR

POWER FROM THE TIDES AND WAVES

D. I. H. BARR

Historical and introductory sketch

Early tide mills
The idea of extracting usable power from the rise and fall of the tides has attracted the attention of ingenious men for centuries. It is possible that water mills, based on small tidal basins or ponds, were in existence on the North Atlantic coasts of Europe by the eleventh century. The earliest tide mills involved a basin in which the high tide was trapped, with release of the impounded water through a passage containing an undershot waterwheel* after the tide level had fallen. Later, breast-shot wheels were also used, and multiple-wheel installations became common. Where a drainage watercourse ran through the pond site, or could be diverted into it, this was used to augment the flow available from tidal storage, and some instances must have occurred of cases intermediate between tide mills and normal water mills.

Without too much modification, such systems continued to be built in favourable locations until the time of the Industrial Revolution. Even thereafter, installations in existence were replaced more or less down to the basic pondage. Less fundamental rebuilding—replacement of wheels and milling components and the like—was commonplace throughout the

* With the undershot wheel, the flow passes below the horizontally-set axle, with little element of containment in the blading. With the breast-shot arrangement, the flow passes under the axle but is introduced at up to axle level and, in conjunction with the containing passage, the blading may so constrain the flow that some of the potential energy of the inflow is recovered (in contrast to the recovery of kinetic energy in the undershot arrangement). The great wheels of the eighteenth and nineteenth centuries were overshot, with the flow introduced at the uppermost level so that it passed over the axle, carried by peripheral buckets, and the consequent reduction in the potential energy of the water flow produced a torque at the axle.

nineteenth century. In 1941 Wailes listed ten tide mills of seventeenth and eighteenth-century origin or reconstruction which were still in operation in Southern England. Relatively basic systems of the same period continued in working order also in France and America. Wailes' survey is of particular interest in showing the continuing use of tidal power at commercial or semi-commercial level until quite recent times. It seems likely that the gap between this late period of the tide mills and the era of preservation of characteristic industrial installations has not been too great for examples to be saved for posterity.

As well as the tidal-basin systems, cases are recorded of waterwheels, either mounted on floating platforms placed in strong tidal streams, or supported from bridges which crossed tidal waterways. A piped water supply based on this arrangement was in operation from Old London Bridge between about 1580 and the removal of the Bridge, with its constricting multiple arches, in 1824. For more standard installations, arrangements were sometimes made to lift and lower the axis of the wheel to suit the varying water levels. Again, inlet controls were devised which allowed breast-shot operation at the periods of greatest difference in water level, and undershot operation in less favourable circumstances. However, for the more common basin-based systems, periods of operation were relatively short—at best about two fifths of the time at periods of spring tides (when the tidal range is a maximum), and reducing to almost zero in some cases during the intermediate neap tides of the fortnightly cycle. Moreover, the periods of running were controlled by the lunar day rather than by the solar day, with which mankind's activities are normally phased. Despite these difficulties, which represent continuing constraints for most potential developments of tidal power, tide mills were a widespread part of life for many centuries.

The Industrial Revolution brings cheaper power
As with non-tidal waterwheels, the onset of the Industrial Revolution in western society meant that cheaper sources of power became available, insofar as completely new installations were concerned. We can only imagine how development of tidal power might have progressed, had there not been such plentiful reserves of coal available for the firing of steam engines. Probably the earliest true hydraulic-model experiments had been conducted by John Smeaton, prior to 1759, on undershot and overshot waterwheels; he understood enough of the scaling principles involved for progress to be made in design. Bernshtein, a Russian protagonist of the use of tidal power for almost four decades, has shown how experience in China during the last three decades provides something of a time-compressed view of the variety of installations which might have succeeded

those of the western pre-industrial era, had the value of mechanical energy relative to man's working time not decreased so dramatically in the west. Many small tidal installations of varying degrees of sophistication have been constructed in China during recent years.

Long-term studies leading to modern concepts
From the foregoing it already emerges that in the development of concepts and their actual application, there are few clear-cut divisions, and few outright breakthroughs, such as have occurred in other fields. This point can be highlighted by a somewhat ironical circumstance. Norman Davey published *Studies in Tidal Power* in 1923, in which he surveyed *inter alia* the potential sites in the United Kingdom. Eight years previously, Davey had published *The Gas Turbine*, dealing with what he forecast as being the internal-combustion engine of the future, although there had not then been any successful demonstration of the principle. While the subject of his earlier protagonism has become a manifest reality, and the critical breakthroughs in its development are well-defined points in recent history, the matter of tidal power has languished in comparison—even although, when Davey wrote, tidal power was being used on a small scale in various parts of the world.

As long ago as 1737, a French engineer, Bélidor, proposed development of the basin configuration (and level control therein) for the existing type of tide mills, which would make them less dependent on the state of the tide. From then on, proposals, patents and desk studies relating to tidal power have followed in an increasing succession, but with intermittent peaks of activity. Thus a wide overlap occurs between the slow abandonment of the traditional tide mills, using waterwheels, and the development of modern concepts based on the use of turbines of varying degree of complexity. Indeed, a triple overlap occurs from the mid-thirties when full-scale experimentation with new turbine types was initiated, particularly in France. This was largely motivated by the desire to be ready to proceed with full-scale construction of these, the key components in modern tidal-power development, when the opportunity arose.

The critical points which do emerge concern the decision (about 1960) to proceed with the commissioning (1966) of the first modern full-scale plant, now in operation on the La Rance estuary in Brittany. This is a single-basin scheme, but with generation of power during the outflowing and inflowing periods, without duplication of equipment—by virtue of the utilization of a turbine type which stems from the pre-war investigations.

Technical considerations
To convey more regarding the constraints governing the design and economics of modern tidal-power schemes which may follow that at La

M. Brigaud, Photothèque EDF

La Rance barrage

Rance, it is first necessary to outline some aspects of the nature of tides and tidal flows, and of the mechanical principles and techniques involved in the various components of such schemes. Of primary importance is the nature of the tidal phenomena at a given site. In smaller schemes, the actual boundaries of the basin have little effect on the conditions at the entrances to the basin. For really large-scale power production, the natural configuration of the tidal inlet will be, in part, responsible for the nature of the tidal variation, and the artificial changes may modify the existing phenomena.

Naturally-occurring basins may be modified in shape, partially enclosed and perhaps divided by artificial embankments. Sluices will probably be necessary to allow inflow of water, and must then be able to be closed. Turbines which will allow generation of power in some form are essential, and may generate in one direction of flow only. However, as already indicated, designs are available which can generate in both directions and also pump in both directions, besides acting as augmenting sluices. In certain possible configurations, independent uni-directional pumps would also be needed. Because vast flows of water will be involved, many sets of turbine pumping and sluice equipment will be required.

For large-scale generation of power, it is most likely that alternators will be coupled directly to turbines, these alternators perhaps also serving as motors in the pumping modes. However, other "linkages" have been proposed, for reasons which will emerge. The power must be led from the individual installations, at least to some nearby location for direct utilization, but more probably into a central grid system. Thus alternating current should probably be regarded as the norm both for the power flow from the individual turbine-alternator sets and for subsequent distribution away from the whole installation, though direct current and other power conveyance methods are possibilities.

Aspects of the above components of potential tidal-power systems will now be considered separately.

The oceanic and nearshore tides

The idealized concept of the equilibrium tides
If a body of the mass of the earth, covered by a uniform depth of water of the same volume as that of the oceans, were to revolve relative to a body of the same mass as the moon at the average distance of the moon (and without rotation other than that necessary to maintain the same orientation along the common axis), as shown in figure 4.1, and if the freezing of the uniform sea could be discounted, small bulges of water would form, culminating at the point closest to the moon and, almost symmetrically, at the furthermost point. There would be a compensating drop in level between these high points, i.e. with the minimum depths on the great circle normal to the earth-moon axis.

In the situation posed, there must be a net balance between the centrifugal forces of revolution of the earth about the centre of gravity of the

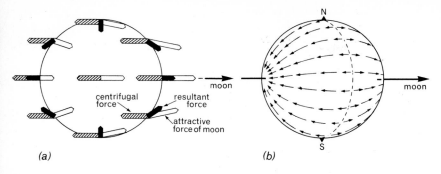

Figure 4.1 Schematic illustration of the tide-producing forces resulting from the earth-moon interaction.
(a) Disposition of resultant forces on elements, on vector combination of attractive and centrifugal forces (after Defant).
(b) Disposition of horizontal components of resultant forces—the tractive forces.

earth-moon system (this lying within the earth) and the summation of gravitational attraction by the moon on all the elements of the earth including the water cover. The gravitation attraction on individual elements varies with distance from the moon, being greatest at the nearest point and least at the furthest point. However, when taken in conjunction with centrifugal force, the net effect is to produce forces which act away from the centre of the earth at the nearest and furthest points from the moon, but towards the centre on the great circle normal to the earth-moon axis (figure 4.1a); they give rise to the horizontal tide-producing components (figure 4.1b) and hence the "equilibrium tide".

Neglecting the reality of the need for water transport, the above equilibrium tide could be imagined to move as the earth rotates, so that the accumulations of water are always nearest to and furthest from the moon. High and low waters each occur twice per lunar day, i.e. approximately every 12 hours 25 minutes.

A basic astronomical relationship with tidal range
By carrying through the same process for the sun, an equilibrium tide of period 12 hours can be calculated to have a range of $1/2 \cdot 17$ of that of the moon. The timing of the tides relates to the transits of the moon above the point of observation, although the average lag of the two tides between each upper transit varies from point to point. This is because the tidal disturbances relating to the moon's effect are dominant.

However, at full moon and new moon, the sun, earth and moon are aligned. This is the condition of *syzygy*, in which the horizontal com-

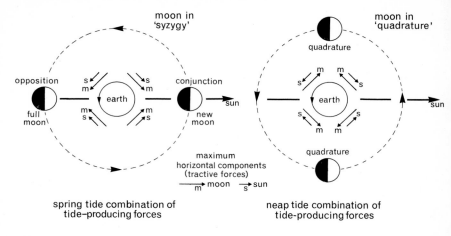

Figure 4.2 Comparison of spring-tide and neap-tide situations, showing basic combinations of moon's and sun's tractive forces.

ponents of the tide-producing forces of the moon and sun act in the same direction. Tides of the greatest range are observed after a delay of a day or so from this combination of forces, as illustrated in figure 4.2. The *quadrature* positions of the moon result in the horizontal components of the lesser tide-producing forces of the sun acting in the opposite direction to those of the moon, and hence give tides of the least range.

The fact that the moon (and the sun) does not move equatorially, but has variable declination, can give rise to variation in tide range between the upper and the lower transit tides, this being known as diurnal inequality.

Tidal conditions for potential power projects
Other things being equal, the requirement for the tidal-power situation is semi-diurnal tides of as great spring-tide range as possible, with little diurnal inequality and without undue decrease in range at neaps. This immediately admits the great rarity of semi-diurnal or diurnal tides in phasing relating to the sun—by which phasing the activities of mankind are normally ordered.

The range of tides along the main continental alignments varies quite considerably. Along much of the Atlantic coasts, especially the North Atlantic, lunar semi-diurnal tides prevail, with relatively little diurnal inequality and with neap ranges not dropping far below half the spring range. Also, there are some prime cases of high tidal range. Regular semi-diurnal tides also occur in the White Sea area and along certain shores of the Indian ocean, but semi-diurnal tides are rarer along the Pacific coasts.

On consideration of an amplitude ratio which is based on the relative importance of individual tidal components combining to produce the observed tides, Defant gives four examples typifying the above outline. Immingham, on the Humber in England, has a high range of spring tides, fair neap tides and little diurnal inequality, and an amplitude ratio value of 0·11. With an amplitude ratio of 0·90, San Francisco has quite moderate spring and neap ranges, with a pronounced trend to the semi-diurnal type of tide at the lowest ranges, and some trend towards that type of tide at the highest ranges. Manila in the Philippines, with a value of 2·15, has a dominantly diurnal type of tide with the semi-diurnal characteristics coming only at intermediate ranges, while Do-Son in the Gulf of Tonkin, with the very high value of 18·9, has a diurnal tide of moderate spring-tide range and minimal neap-tide range.

In favourable circumstances, conditions for tidal-power schemes may exist in respect of basins feeding directly to the main continental alignment. However, these alignments are often broken by channels—particularly converging channels—in which the tidal effects are magnified, either by convergence or by resonance, or by both. It is here that the prime opportunities for tidal-power developments occur, either by the utilization of parts of such channels or basins connected to them.

Turbine and pump units

Principle of change of angular momentum of water flow

The objective of extracting the potential energy available from the flow of water from a higher surface level to a lower level is accomplished as follows. The flow is given a rotation about the axis of the turbine by stationary guide vanes—a tangential component of velocity. Then the flow passes on to the moving blades of the runner which are set to align into the moving stream in a streamlined manner (figure 4.3). This implies that the correct combination of tangential and throughput velocity of the water, and tangential velocity of the leading edge of the blades all obtain. In its passage through the zone of influence of the moving blades, the tangential component of velocity of the flow must be removed by the path imposed upon the flow by the blade shape, so that the flow leaves the trailing edge without a tangential component.

To remove the tangential component, a torque must be exerted on the water by the moving blading of the rotor, and conversely on the blading

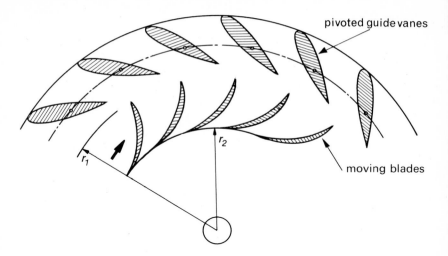

Figure 4.3 Flow path through turbine, redrawn after Daugherty and Franzini. As is normal practice, this is drawn for two-dimensional flow through the moving blades, i.e. through-flow is radial inwards. For the turbines discussed, three-dimensional representation would be necessary, although the principles involved are the same.

by the water. This torque drives the alternator, which converts the mechanical energy to electrical energy. At the trailing edge, the flow still has a comparatively high kinetic-energy content, but the presence of the draught tube—a gradual expansion of the flow cross-section—allows the recovery of a large proportion of the kinetic energy. At the point of leaving the overall flow tube through which it is led, the velocity is small and the kinetic-energy content is low. This "recovery" of kinetic energy is inherent in the overall design of the turbine passages and blading for a particular duty—a combination of rotational speed, head difference in terms of the fluid flowing, and throughflow or discharge, which is considered as the design condition. The net result is that with well-designed components, about 90% of the potential energy of the difference in level between almost quiescent bodies of water close to the turbines can be converted, when running under design conditions.

Constant-speed running for alternating current
High efficiency can be maintained with change of head difference if the rotational speed is governed to be proportional to the square root of that difference, with the power output then proportional to the 1·5 power of the head difference. However, if the objective is to feed alternating

current to a grid system, using a directly-coupled or fixed-gear coupled alternator, the rotational speed of the turbine is fixed by matching the alternator construction with the frequency of the grid. As the head available decreases, the efficiency can be maintained to some extent by adjusting the angle of the inlet vanes to increase the tangential component of velocity in comparison with the throughput component. A more significant step in maintaining efficiency over a range of head difference, while retaining a fixed rotational speed, is taken when the blade setting of the rotor is made variable. The name Kaplan is almost universally applied to vertical-shaft propeller turbines with variable-pitch

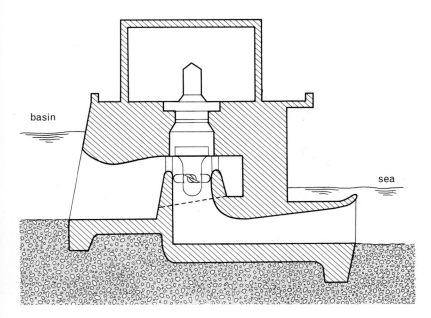

Figure 4.4 Conventional vertical-access Kaplan turbine applied to tidal-barrage situation.

blading, and with throughput component of flow at entry to the moving blading almost completely axial (i.e. vertical downwards). The tapering inflow passage, called the spiral casing, is led round the turbine runner (moving-blade) axis, with complete peripheral feed to the inflow guide vanes (figure 4.4).

Requirement of submergence to prevent cavitation
The high velocities occurring close to the moving blades result in pressure

reductions, and potentially to vapour cavities forming where the absolute pressure drops below vapour pressure. Such *cavitation* is undesirable because the subsequent collapse of the cavities upon re-pressurization could result in erosion damage to the blades. Cavitation can be prevented by placing the turbines at an adequate depth below tail-water level.

In hydro-electric practice, large-diameter turbines are sometimes sited in low-head schemes. A prime justification for large size is the reduction of the number of units, which tends to economies in the corresponding ancillary control and generation equipment. In other situations, equally powerful units of smaller diameter are sited deep below the lower water-surface level of pumped-storage schemes. Submergences of 50 m or more are common, because these represent the best economic compromise.

It could be expected that a conventional propeller turbine, with or without variable pitch, could operate in the reverse-direction pumping mode, with a best efficiency of the order of 80%, provided that the spiral casing and the like were designed with this end in view.

The position was thus set for the step forward to turbines which could be particularly suitable for tidal-scheme conditions.

Recent turbine developments

For decades there have been proposals for turbine arrangements where the throughflow component of velocity is totally axial, or more nearly so than applies with the conventional vertical-shaft arrangement. Thus, even while the development of the concept of the totally axial throughflow component of velocity on to the runner (after a radial guide-vane stage) was proceeding to its present advanced state, pioneering arrangements to transform the guide-vane throughflow component from radial to axial were being tested and installed. Three arrangements seem to hold potential for tidal-power applications. The first of these is already in use.

Bulb turbines

In the bulb arrangement, the generator-motor unit is contained within the flow passage. Figure 4.5 shows the general proportions and arrangement of the units which are installed at the La Rance tidal-power scheme in Brittany. These units are designed for 5-mode operation, as defined in Table 4.1, and have the maximum efficiencies shown there.

A great deal of experimentation followed the successful use of single-mode bulb turbines for small hydro-electric installations. At St. Malo, a pilot installation was operated and, in addition, many different arrangements were tested at model scales. In figure 4.5 the adjustable guide vanes are upstream of the runner for the primary generation flow, as is conventional in single-mode turbines. Reverse-flow generation involves

POWER FROM THE TIDES AND WAVES

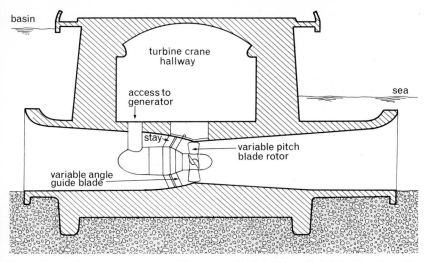

Figure 4.5 La Rance configuration of barrage and bulb turbines. In this turbine alone, the generator is within the water passage.

Table 4.1 Operating modes and typical performance data for La Rance bulb turbine (single unit).

Operating mode	Head difference (m) sea low +ve	Flow ($m^3 s^{-1}$)	Power (MW)	Efficiency (%)
Generating— flow seaward	11	100	9·0	85
	7·8	136	9·0	86·5 (max)
	5·3	250	9·0	69
	5·3	131	5·4	80·5
Generating— flow landward	−11	120	9·0	71
	−9·7	135	9·0	73 (max)
	−7·2	227	9·0	54
	−7·2	116	4·7	71
Pumping— flow seaward (virtually abandoned)	−5·2	120	−9·0	67·5 (max)
	1	300	−4·0	—
Pumping— flow landward	4·4	85	−6·0	64 (max)
	−2	300	−8·0	—
Sluice flow seaward or landward	186	—	—	±1
	322	—	—	±3

imparting a tangential component of velocity by the runner blading, which thereby receives torque, and the recovery of the rotational kinetic energy of the water by the removal of the tangential component at the guide vanes. This was accepted as the best compromise in the circumstances. To maintain the ideal of having the runner downstream of the guide vanes in generation and upstream in pumping, while allowing four-mode operation, it is possible to provide:

(a) guide vanes on both sides of the runner, with the set of vanes which is not required for a particular mode of use set axially.
(b) dual runners, both with variable-pitch blading, one of which acts as stationary guide in any given mode of flow.

Table 4.2 Trends in non-dimensional size and non-dimensional specific speed (rev) of different turbine designs.

Runner diameter (m)	Design head (m)	Output (MW)	Speed (rpm)	Non-dimensional diameter	Non-dimensional Specific speed (rev)	Details of turbine
1·23	313·3	29	600	2·97	0·0745	High-head Francis
3·38	27·4	25	112·5	1·42	0·272	Low-head Francis
7·93	28·3	758	90	1·37	0·525	Modern Kaplan (6-blade)
				4·4	0·11	Modern pump-turbine (Francis)
5·35	7·8	9	94	1·46	0·657	La Rance bulb
7·2	7·5	17	60	1·39	0·605	Large bulb (design study)
3·3	10·25	6·3	125	1·32	0·519	Early peripheral
12·2	6·1	37·2	38·6	1·36	0·75	Large peripheral (design study)

Neither of these is as immediately attractive as the arrangement in figure 4.5 with the bulb upstream in the primary generation mode, but the situation described in (b) has been implemented in a pilot hydroelectric installation on the River Truyère in France, with the generator bulb downstream of the alternating runner/fixed blading pair of rotors in the primary generation mode.

In the different bulb arrangements—as in the two other basic possibilities, peripheral generator turbines (or straight-flow turbines) and S-tube turbines—supporting members must obstruct the flow, causing some energy losses, however well streamlined.

Trends in non-dimensional size and non-dimensional specific speeds of different turbine designs are shown in Table 4.2.

Peripheral-generator straight-flow turbines

The peripheral-generator arrangement was first proposed by an American engineer, L. F. Harza, in 1919. Figure 4.6 shows its basic configuration. Many examples of this type of turbine were installed in low-head hydro-electric schemes on the rivers Iller and Lech in Germany, and also in Austria. After difficulties with sealing the long peripheral joints, these are now running successfully. More recently the English Electric Co. (now integrated into the General Electric Co.) have devoted considerable attention to the problem of sealing, with tidal installations in mind. Early

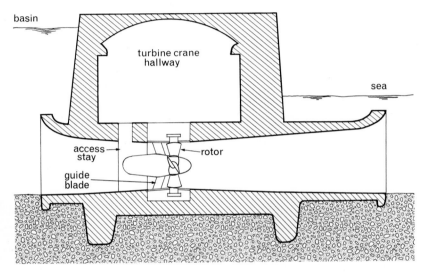

Figure 4.6 Peripheral-generator straight-flow turbine in tidal-barrage context. The generator surrounds the water passage.

in 1976, the Swiss firm of Escher-Wyss announced the availability of a new range of peripheral-generator turbines under the trade name "Straflo". These can be of the reversible-flow type, and have adjustable-angle runner blades. There is stated to be no limitation on size. New methods have been developed to overcome the problem of sealing, which had plagued the earlier installations.

S-tube turbines

In contrast, the practical difficulties involved in sealing the S-tube type of turbine are minimal (figure 4.7). Its hydraulic characteristics and those of the bulb turbine are, however, essentially less favourable than in the case of the straight-flow turbine. A number of small installations exist

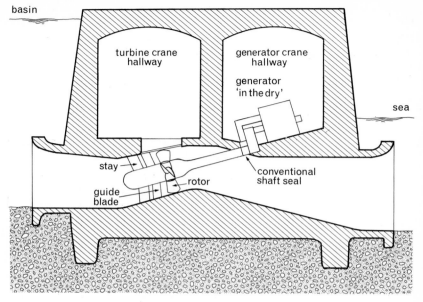

Figure 4.7 S-tube turbine in tidal-barrage context. The generator is completely clear of the water passage.

in low-head hydro-electric schemes in America, and this type of turbine is being developed by the American turbine-manufacturers Allis-Chalmers.

Other components

Alternators, dynamos or other energy conversion methods
Production of electrical energy by alternator, directly coupled to the turbine, is the starting point in consideration of potential tidal-power schemes. Most electrical power is produced in the form of alternating current, and all central grid systems are so based. It has also been demonstrated that the constraint of constant-speed running has a profound effect on the complexity and cost of the turbo-machinery.

Once consideration is given to a departure from the normal practice—the practice in which past experience has been built up—many possibilities emerge. If an industrial process using direct current could be integrated with the tidal-power scheme, the need for constant-speed variable-pitch turbo-machinery would be obviated—but the flexibility inherent in normal practice would be lost. As with hydro-electric installations, one of the great attractions of tidal-power schemes is that, once the outlay has been made on capital cost, running costs are small and the life should be extremely long.

The basis of economic practice in the Western World is that the capital outlay should be recovered by an amortization scheme extending over a conservative estimate of the life of the plant. If the industrial process to which a power scheme is linked were to become less economically important, great expenditure would be necessary to re-arrange the tidal scheme for more-general usage.

There is also the question of breaking the constant-speed constraint, while maintaining the end-product of alternating current suitable for feeding into a central grid. There are various possibilities, all of which would at present appear to require increased capital outlay. There are also development difficulties, but these have been overcome in the analogous experience of the direct production of alternating current. Direct current could be produced and converted to alternating current. It has also been proposed that direct-action pumps could be directly coupled to the turbines to produce a reservoir of high-pressure fluid which would, in turn, feed a Pelton-wheel turbine* driving a normal alternator.

This latter type of turbine is very efficient over a wide range of loadings while running at a constant speed, provided that the head available is constant. The load variation is accomplished by modifying the flow rate of the fluid by a needle valve in the nozzles which direct the jets. The fluid is recycled.

Another possibility is the use of direct current for the production of hydrogen. Much has been written about the hydrogen economy in recent years, and the pipe flow from the tidal-power scheme in the hydrogen mode could be either to a land-based power conversion system or could be taken further. The latter suggestion is a basic aspect put forward by the protagonists of the hydrogen economy.

Sluices

The need to allow free through-flow of water at times of filling of basins is obvious. Depending on the circumstances, it could be possible to arrange that closures and openings can be made with equal levels on either side. Again, if filling can be accomplished without too great a change in level through the sluiceway, there will be less danger of scour.

However, in any large scheme, sluices will be considerable structures and, at the limit, it has been suggested that sluices are inefficient and that pumps should be used instead. Pumps will obviously use power,

---

* In the Pelton wheel, a tangential jet impinges on specially-shaped buckets on the periphery of a moving wheel, with the bucket velocity about half the velocity of the jet. The fluid is virtually brought to rest in passing over the curved bucket shape, and falls away just clear of the wheel.

and this may seem contrary to the whole concept of power production from the tides. Within the constraints resulting from the tidal periods, it may be more efficacious to pump against a very small head (and thereby be able to have more stored water available for use in generation when the head is large during the generating cycle) than to allow filling merely by gravity flow through openings.

Active barrage framework and float-in proposals
The *active portion* of the barrage refers to the portion through which water will be passed (other than by leakage). This may be either a machine passage or a sluice.

In the La Rance scheme, much of the total expenditure went on a comprehensive cofferdam arrangement, i.e. a temporary, watertight barrier around the actual construction location within which the barrage could be constructed under dry conditions after the enclosure had been sealed and pumped out—very much in the way that hydro-electric barrages have normally been constructed. (Because of the low heads involved in tidal power, the proportion of such items as barrage-containing machinery, access galleries, sluices and controls is likely to be much larger than in hydro-electric practice.) The site was favourable, and the greater part of its length was needed for the active barrage arrangements.

Since the construction of the La Rance scheme, experience gained in other fields is considered to be highly relevant to the construction possibilities, and hence to the economic feasibility of tidal-power schemes in the future. Essentially this is that the active parts of the barrage should be constructed in assembly-line manner at some convenient point on the shoreline, and should thereafter be floated into position and sunk in the manner of an oil production platform. On the one hand, difficulties regarding depth, water conditions and the like are not of the same magnitude as those in the case of oil production platforms, but the latter are never placed in a position of closure, or near closure, where the continuing tidal flows will tend to dislodge them until the closure is completed.

Much effort in the design level has gone into the planning of assembly-line-built float-in components for tidal-power-scheme barrages. Indeed, the second full-size tidal-power scheme to be built employed this system. This scheme was at Kislaya Guba, north of Murmansk in the USSR and was completed in 1969. The engineer in charge was Bernshtein. In a particularly favourable site as regards narrowness of the natural opening into an embayment, it was possible to arrange that one float-in section virtually completed the closure. This scheme is regarded as a pilot scheme. On the other hand, the tidal rise and fall, which can be greater or smaller depending on the favourability of the site, does not represent a variable in the sense that basin size is a variable and, even although only one unit has been used, the whole can be regarded as representing a section of a potentially larger system rather than a scaled-down model. Further details of this scheme are given on p. 125.

Non-active barrage construction

As will again emerge in the review of schemes and proposals, the non-active barrage component may be minimal in particularly favourable sites from the point of view of the shore configuration, and where it is accepted that the production of energy is the dominant economic factor. On the other hand, in proposals relating to the production of firm power, not tied to the lunar day and the alignment of the moon with the earth-sun axis, the construction of the non-active barrage, or dyke, may become a very significant part of the whole.

Again, experience is continually being gained in the economic use of different materials for the construction of water-retaining barrages in analogous circumstances. This can arise from storage-reservoir practice, from coastal impoundments for the storage of water, and from the early stages in the construction of coastal impoundments for reclamation of land. Such reclamation can involve final raising of the land level above maximum sea-levels or may, as in the case of considerable areas in Holland, involve construction of embankments which must be more or less impervious.

In the situation of the tidal-power scheme, embankments must be able to withstand considerable wave action, but not necessarily wave action associated with fully exposed coasts.

Systems have been devised for the use of rock-fill with central impervious membranes, and for the use of very much finer sand-fill, with less obvious tendency to direct percolation and where the watertight membrane may be either central or closer to the surface and may be partly associated with protection against wave erosion. Centrally located clay or even light sheet-pile membranes are possibilities. Recent experience in Holland has led to the development of low-viscosity bitumen compounds which can be run into fill material to provide the foundation of a wave-resisting facing to exposed dykes. This form of treatment may also provide an alternative means of forming internal watertight zones.

The economic and environmental implications of large-scale barrage construction would have a considerable influence on the viability of some of the more-sophisticated schemes now being considered.

Methods of operation and integration with power distribution system

Single-basin uni-directional power production

The simplest arrangement for a modern tidal-power installation is to repeat in modern large-scale terms the arrangements of the basic tide mill. A single basin is formed by a closure across a tidal inlet, in which are placed a number of turbines with associated electrical-generation

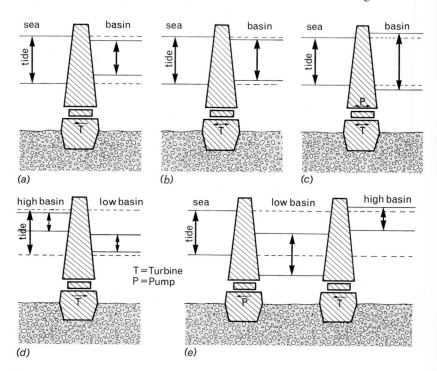

Figure 4.8 Possible ranges of level variation in basins, as compared with seaward tidal range. T indicates turbine flow; P indicates pump flow. Sluice flows and pump-aided sluice flows not shown.

equipment. There are also sluices in the barrage to allow inflow of the rising tide without undue impediment. Certain types of turbine design may be utilized for this purpose, in combination with supplementary sluices. At high tide, or soon after, the sluices and (if appropriate) the turbine inlets are shut. Seaward of the barrage, the water level falls and, when the drop in level reaches perhaps half the tidal range or somewhat less, the control gates leading to the turbines are opened. Sufficient

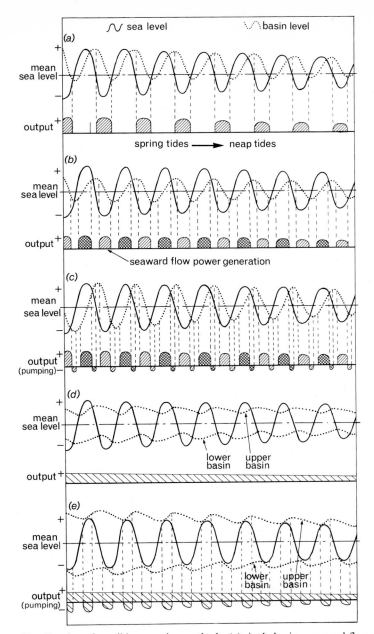

Figure 4.9 Sketches of possible operation methods: (*a*) single basin—seaward-flow power generation; (*b*) single basin—seaward and landward-flow power generation; (*c*) single basin—seaward and landward-flow power generation and pumping; (*d*) double basin—power generation between basins without pumping; (*e*) double basin—with pumping between sea and basins.

turbines are provided to utilize the maximum rate of fall on average tides, and power production is initially more or less at a constant head (figure 4.8a and figure 4.9a). Power production continues until it is rendered inappropriate by an insufficient head difference due to the rising of the tide again, and the sluices are then opened so that the cycle repeats. Fixed-blade turbines are a possibility, with variable-pitch blading giving greater output if alternating current is to be produced, but at additional capital cost. Were the power to be utilized in a direct process, it might be inappropriate to consider elaboration—even of variable-pitch blading—but, if the power is to be fed to a central system, there is at least the need to examine whether a benefit would emerge from the additional investment.

Single-basin two-directional power production
The production of power on the inflowing cycle is a fairly obvious development (figure 4.8b and figure 4.9b). This aim is achieved in the only existing full-scale tidal-power scheme on the La Rance estuary, and the arrangement for a turbine/alternator device, doubling as a pump/motor, is necessarily more complex (figure 4.8c and figure 4.9c). Power production is still phased with the (lunar) semi-diurnal tides.

Double-basin uni-directional power production
The total available basin area is split in two, with uni-directional turbines set in the dividing barrage (figure 4.8d and figure 4.9d). The flow is from the higher-level basin, intermittently filled up at the end of the rising-level period, and into the lower-level basin, intermittently drained during the last stages of falling tide. If the output requirement is considerably reduced, a power-production cycle which accords with solar time can be achieved. The term "firm" is applied to the capacity to generate power when required, and with the two-basin arrangement a considerable increase in degree of "firmness" is achieved.

Other basin arrangements
Desk studies of favourable sites have led to proposals for more-complicated arrangements involving three basins, with the objective of achieving greater firm-power output, with reasonable economy in installation costs.

Pumping to augment output, and matching of pumped storage with tidal-power production
Mention has already been made of the possibility of augmenting power output by pumping. In the La Rance estuary tidal-power scheme, pumping is possible in both directions although, in practice, pumping has been found to be economic mainly in the landward direction. If

additional water can be stored by pumping against the small head at the peak of the tide, and energy recovered when the head itself is much greater, there can be a net gain in energy. This capacity may be particularly valuable during the neap-tide cycle, when the creation of additional storage does not cause flooding above the natural high-water level in the basin. There may also be advantage in providing during the operating cycle power output which is more commensurate with that available during the spring-tide range.

Again, there are no clear-cut divisions between augmenting power output by pumping, creating pumped-storage capacity, and using the pumped-storage concept to balance power output. At the other end of the scale is the classic situation—which, however, may never arise in practice. At an ideal site for tidal-power production, a simple unidirectional scheme provides the maximum of power at the minimum of cost. Alongside the favourable configuration for tidal-power production is a favourable natural configuration for pumped storage, with a good high-level reservoir site close to a plentiful supply of low-level water. A standard pumped-storage installation is provided with pumping power by the excess of energy production from the tidal-power scheme over the steady demand. Whenever the power production falls below the demand, and during the comparatively long periods of zero production, the steady demand is satisfied by the pumped-storage scheme in the production mode.

Reality will always be vastly more complicated than this, but considerable development of the pumped-storage concept has taken place and is taking place in the United Kingdom at sites which are not too distant from potential tidal-power sites. Superficially, tidal power, if produced at a sufficiently low cost, could be absorbed by existing pumped-storage schemes, and around 70% of the power recovered when the demand is high. However, it is unlikely that pumped-storage capacity will outstrip the need for its use to balance the output from nuclear stations, where continuous uniform power production is most desirable, and to provide fast-availability standby capacity. In fact, even in an island like Britain, with a long coastline and some prime tidal conditions, the potential production of tidal power at reasonable capital cost is not excessive compared with current demand. Thus tidal-power production could probably be taken into the system without undue difficulty, with the total capacity still largely met by fossil-fuelled power stations, provided that the economic balance was favourable. While not corresponding directly with figure 4.8, figure 4.9 shows possible patterns of power production and of pumping-power absorption for different methods of operation.

Installations and projects

To illustrate the varying manner in which solutions may be proposed within the constraints which have been outlined, a number of projects have been selected from the literature. The list is far from exhaustive. It has already been indicated that in the Western world at the time of writing, only one full scheme (La Rance) has been constructed; none are under construction, although preliminary moves may have been made in the USSR, in addition to the construction of one pilot scheme.

The La Rance estuary scheme receives less attention here merely because considerable reference has already been made to it as the only full-scale modern scheme.

La Rance estuary

This is a single-basin five-mode-machine scheme, which was fully commissioned by 1969. The 24 bulb configuration turbine/pumps each have a capacity of 10 MW, and the net energy production is now in the region of 500 GWh per year. The barrage is situated quite near the mouth of the 21 km estuary, in which the average spring-tide range is about 11·5 m and the average neap-tide range 5·4 m, with predicted extremes of 13·5 m and 3 m (figure 4.10). The tides are semi-diurnal with minor diurnal inequality. Peak flows at the barrage site had been found

Figure 4.10 La Rance estuary with barrage site.

to range up to 20,000 cubic metres per second. The active section of the barrage, the major part of the closing length, was completed under cofferdam conditions.

Kislaya Guba experimental tidal-power plant
A particularly favourable basin configuration for a pilot project was adopted, despite comparative lack of tidal range (from 1·3 to 3·9 m) whereas, in other regions of the White Sea area, ranges of 10 m obtain. Semi-diurnal tides operate a single 400-kW 5-mode-flow bulb machine of French manufacture, with provision of conduits for later installation of a Soviet-built machine. The surface area of the basin is 1·14 km^2 and the installation is intended essentially to provide experience. As already mentioned, the active barrage comprised one float-in structure, which completed the closure in one operation.

Severn Estuary and Bristol Channel
This is the classic site in the United Kingdom, with the semi-diurnal tides of average spring range 11·6 m, and average neap range 5·8 m in the region of the early barrage proposals landward of Bristol. Davey outlines the 1920 scheme of the Civil Engineering Department of the Ministry of Transport. This included a linked pumped-storage scheme some miles inland and a railway crossing, together with a ship lock. A single-basin uni-directional power-generation system was envisaged, with direct-current dynamos connected to the turbines. Thereafter, interest was maintained, with notable studies utilizing tidal models undertaken by Professor A. H. Gibson at Manchester University during the late nineteen-twenties. Official reports and private proposals have appeared at intervals, and recent detailed studies have been undertaken by the Central Electricity Generating Board.

Dr. T. L. Shaw had proposed various double-basin arrangements, with the smaller basin sited in deeper water, preferably without any natural coastline boundary, and able to be pumped to well below normal low water. Thus a linked tidal-power pumped-storage system would be created (figure 4.11a).

The Central Electricity Generating Board concept, however, returns to a somewhat more conventional double-basin layout, with the possibility of the low-level basin being isolated from the shoreline (figure 4.11b and c and figure 4.8d). Depending on the layout and the extent of utilization in a pumped-storage mode, firm daytime power output throughout the spring-neap cycle might be raised as high as 5000 MW, with a net output

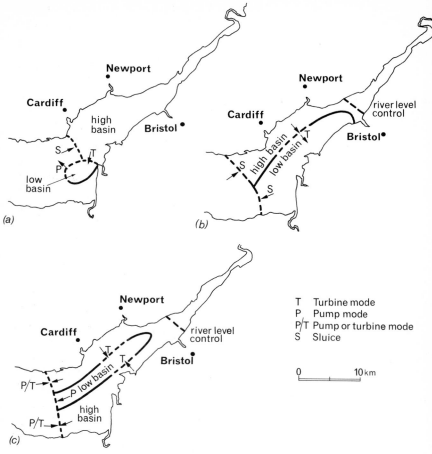

Figure 4.11 Layouts for possible development of Severn estuary.
(a) Shaw layout—basin arrangement as in figure 4.8e.
(b) CEGB layout—basin arrangement as in figure 4.8b.
(c) CEGB layout—basin arrangement as in figure 4.8d or e.

from the tidal-power component of the system about 20 TWh per year.*
At the other end of the range of possibilities, about 25 TWh net output could be obtained from a continuously running system, but where the output varied between 3500 MW and 2000 MW with the spring to neap cycle. For comparison, the 25 TWh figure is of the order of 13% of the

* TWh = terawatt hours = 10^{12} watt hours
  GWh = gigawatt hours = 10^9 watt hours
  MWh = megawatt hours = 10^6 watt hours

Central Electricity Generating Board (CEGB) annual output in the mid-seventies.

In the CEGB proposals, some pump-turbines could be required, but the principal output system would consist of uni-directional turbines between the high-level and low-level basins. Another point of interest is that at the Severn-Bristol Channel site, significant modification to the tidal ranges would occur as the barrage position was moved seaward.

Bay of Fundy

The Bay of Fundy on the eastern coastline of North America (and, in part, forming the division of the United States and Canada) is noted for the greatest tidal range in the world. Various inlets suitable for major tidal-power projects exist at the head and, to a lesser extent, along its length. Quite regular semi-diurnal tides have overall maximum range at the head of about 16 m, with neap range about 7·3 m. By the southerly junction with the open ocean, the maximum range has decreased to 6·7 m.

Passamaquoddy Bay. Quite close to the entry to the Bay of Fundy and thus not at the location of maximum tidal range (Passamaquoddy entry, about 2·4 m spring range and 1·0 m neap range), there is an extremely complex inlet with the United States/Canadian border running through it. Projects have been planned using this area, and the configuration lends naturally to double-basin arrangements with many possible alternatives. In the 1930s work was actually begun on a tidal-power project (largely in the United States zone) and some small dykes were built, but the matter thereafter lapsed.

Full development of the Passamaquoddy situation requires a combined United States/Canadian project, and recently there has been considerable investment in project planning and design.

Head of Bay projects. In studies during the 1960s twenty or more possible arrangements for utilizing the large subsidiary inlets at the head of the Bay of Fundy were undertaken. As already indicated, tidal range is extreme throughout the area. Both single-basin and double-basin arrangements were examined, and it was concluded that a single-basin unidirectional-flow development operated in conjunction with pumped storage would be the most economically viable proposition in present economic circumstances.

Cook Inlet

The Alaskan inlet in which the range of the tides was noted by Captain James Cook provided another example of the possibility of a simple form of tidal-power scheme linked with pumped storage as the best short-term economic proposition.

The whole inlet is some 200 miles long and, while the entrance tides have a mean range of about 44 m, the range at the head is 7·6 m. There is a pronounced diurnal inequality. Professor E. M. Wilson has outlined proposals for barrages across the smaller arms at the head of the inlet, and the use of very-large-diameter peripheral-generator fixed-blade turbines to gain energy at the least possible capital cost. The corollary is that for normal usage a linked pumped-storage scheme must be provided, and the topography in the vicinity of the inlet must allow an economical pumped-storage development.

San José Gulf
Although the tidal range along this part of the Argentine shoreline is not too great (from 3·5 to 7·8 m), the vast area of the Gulf (780 km^2) suggests its suitability for tidal-power development. H. E. Fentzloff has outlined a development based on peripheral-generator turbines of large size, in this case with power production in both flow directions.

Other sites and more fundamental developments
Possibilities exist for tidal-power projects along certain of the shorelines of the Soviet Union, India and Bangladesh, and in the Far East and Australia, as well as for much larger developments than have already been discussed, again at the locations of maximum tidal range and admitting considerable environmental change.

Assessment

Economic assessments of tidal power are normally made with a degree of enthusiasm, but even the enthusiasts cannot at the present time claim that the case can be made for the economic production of low-grade power by the harnessing of the tides. Electrical power is manifestly a most valuable resource of civilization as we know it, and conveniently provides high-grade lighting, powers sophisticated computational arrangements and precision machinery, and provides convenient power for lifting operations and the like. There is no known substitute. However, a fair proportion of the electric power produced in western society is for relatively low-grade use. We are fortunate to be able to have electric power in a fairly competitive position for the heating of reasonably insulated houses, but this level of abundance of potentially high-grade power must mitigate against tidal-power developments as conventionally costed.

Another way of looking at this is to return to the case of Davey and to compare the present states of the gas turbine and tidal power. A society which can use energy in the jet transport era in the way that western society now does, does not need tidal power by conventional economic criteria. Nevertheless, the basis of future cheap electric power is accepted as based on nuclear fuelling. The long-term hazards of nuclear wastes must be in the minds of all thinking people. A tidal-power scheme, once built, produces no wastes and provides energy for an indefinite period at little additional cost. But so far as the United Kingdom is concerned, tidal power cannot do more than contribute to the total demand for electric power. It is possible to plan tidal-power developments which integrate with improvements in transportation and in recreational use of waterways. Perhaps most important, it would be possible to utilize a great deal of low-grade and intermediate skill in the construction of tidal-power plants.

It is certainly instructive to consider the two circumstances in which existing tidal-power plants in western society have been built. In the case of the La Rance estuary, French technology had planned for the construction of a tidal-power station over many years. French political will decided that a power station would be constructed using French technology, to obtain knowledge of potential value in the future, rather than from a belief that this was a means to cheap power at the present time. Then the Soviet Union power plant was built with one French turbine and a space for a Russian turbine. Some few years later, it has been possible for Russia to provide the lowest tenders for bulb-type turbines for a particular non-tidal installation in Canada, and therefore to obtain the order.

The total potential capacity for feasible, if currently uneconomic, production of tidal energy throughout the world has been estimated as between 300 and 450 TWh per annum. With a mid-1970s annual consumption in the United Kingdom of the order of 200 TWh, the matter is put into perspective. Western society is feeling many strains now, not least those of changing from being a growth society to a possibly stable society without disintegration. It is possible that the construction of certain large tidal-power schemes, if undertaken for the correct reasons and managed in the correct ways, could immediately form a part of this process as far as individual countries are concerned. In the future, if it is forced upon us that electrical energy be used less for low-grade purposes and more for purposes for which it alone is suitable, tidal-power schemes could come into their own because the economic balance would have changed. These remarks are intended to demonstrate the complexities of the situation and in no way suggest that clearcut conclusions can yet be formulated.

Power from waves

Introductory

As already indicated, estimates of the potential annual energy recoverable from tidal sources throughout the world, by reasonable if not currently economic capital outlay, have varied. Professor E. M. Wilson has estimated a maximum of 350 TWh, whereas the United Kingdom annual consumption of electrical energy in the mid-seventies is almost 200 TWh. The position regarding the potential of extraction of energy from ocean waves is quite different. With an average power availability of at least 75 kW per metre of length at exposed offshore locations facing westward into the Atlantic, if an overall efficiency of 30% in extraction of energy and its transportation could be achieved, the total length of wave-energy extracting structure required to approach the current United Kingdom consumption is of the order of 1000 km. This is the basic situation which has led to some considerable interest and investment in feasibility studies within the United Kingdom.

S. H. Salter of the University of Edinburgh has devised a shape which, when floating almost submerged and partially constrained from rocking, can transform a considerable portion of the energy from a train of deep-water waves reaching it, to produce torque. In the ideal circumstances of a flume test, an energy recovery rate of about 90% has been achieved (figure 4.12).

The terms "rockers" and "nodding ducks" had been applied by Salter and his colleagues to systems based on the use of the Salter shape. The term "Salter duck" has now arisen, as a convenient and descriptive name when comparisons are being made with other systems. At the time of writing, comparative studies of four possible systems are planned, but the Salter concept first caught the imagination because of its basic elegance.

It is immediately apparent that, whereas the direct production of alternating currents is the first option in the case of tidal power, it may be less relevant in the case of the capturing of wave power. All that has been said already about indirect linkages in the tidal-power connotation is relevant. Hand in hand with the laboratory flume and wave-tank tests that have so far been undertaken has gone the planning of means of channelling and converting the torque energy and bringing it to land.

Thus, on the one hand, there is for the United Kingdom the potential of a new energy source which would make a significant contribution to energy needs, if this could be carried through a prototype stage to the building of sizeable lengths of full-scale units in an economic manner. On the other hand, the technical problems of development are far in excess of those related to tidal-power schemes, where economic factors are now the main constraint.

Historical

As in the case of tides, the challenge to find a means whereby the energy of the waves can be diverted to useful purposes has aroused the attention of engineers and scientists for many years. As with tidal power, many patents have been registered, but there have not been any large-scale installations which would correspond to the long-standing practical use of tidal power. On a small scale, there has been the use of the rise and fall of a buoy against its mooring to provide a minimum of power for the operation of a recording or transmission system, where the buoy has itself served as a wave-recording instrument, or where a float has been sited near an isolated bed-fixed installation.

Again, and more recently, power has been extracted on a small scale from a hollow partially-submerged float in which water-column oscillations are induced by the waves and the shape of the submerged inlet, and are partially constrained by the egress and inlet of the air above the column being taken through turbine blading. The scaling-up of such devices, originally investigated in Japan, is to form an important area of study.

The Salter shape—the "nodding duck" system

An evaluation of the potential power availability from a train of deep-water waves is given by calculating the "cut and fill" potential energy of a single wave cycle, were there to be a 100% efficient return to mean level. This is available once per wave period, assuming a repeating sequence of waves, and the power may thus be calculated. For real sea conditions, adjustments have been made which allow the significant wave height to be included in the calculation (this being the mean of the highest third of the waves) and still provide a fair estimate of the average potential power. It is this type of calculation that has led to the figure of more than 75 kW for the average power transmission from wave action per metre of open Atlantic-facing front.

Then with knowledge of the long search for a break-through in the matter of wave power, and perhaps with some lessons gained from recent studies in floating-breakwater design, Salter sought to devise a shape which would absorb power efficiently from a deep-water wave. A deep-water wave is a wave where the water disturbance at the seabed is minimal. Salter realized that success must envisage the almost complete stilling of the oncoming wave. Since a rotary motion was to be imposed on the floating system by the action of stilling the wave, the lee side of the device had to be essentially circular in shape: otherwise it would again generate a wave. He obtained a shape for the wave-facing side of the device which would rise with the oncoming wave, thus rotating the whole, and in which the motion of the boundary shape would be compatible with the motion of the water. The result is illustrated in figure 4.12. In initial laboratory tests of the device, the constraining force was applied by an external lever system. In practice, the constraining force would have to be supplied by the resistance of individual Salter-shape floats to rotation about some common continuous axis in which the power-converting devices would probably have to be housed (figure 4.13). The proposals have at times included that a balancing float should lie in the still water behind the device, with cantilever arms at intervals providing support to the central core on which, in turn, the Salter-shape elements would rotate. Consideration has also been given to the possibility of the whole being anchored or, alternatively, being allowed to float free, slowly impelled landward by the waves. This would not unduly reduce the efficiency of the device. Sections would have to be towed out to sea in a cycle of operations. This possibility does not seem too convincing because of the danger of error in changing weather conditions leading to grounding and consequent damage.

In providing resistance from the common core or spine to the torque to be derived from the Salter-shape elements, it would be necessary to ensure that different parts of the length of the unit would always be affected by different stages of the incident wave cycle at one time, so that the torques were balanced. A unit composed of a series of individual "Salter duck" elements would therefore have to be at least one kilometre long.

Some flexibility in the spine of a unit would be necessary at all times, and additional capacity to yield during storms would be essential to prevent "hogging" failure, as has occurred in certain cases with conventional craft. Thus the design of effective coupling and communication systems between individual elements is a fundamental aspect of the development.

Other systems and comparative testing programme
The elegance of the Salter-shape concept perhaps initially drew attention away from other current experimentation. In May 1976 it was announced by the United Kingdom Department of Energy that a Wave Energy Steering Committee had been set up and had selected four lines of investigation for a two-year feasibility study, to be undertaken jointly by the various originators of the proposals and by other relevant agencies. Together with the "Salter duck" system and the development of the

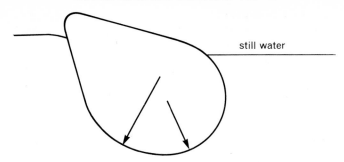

Figure 4.12 Salter cam shape shown without any reactive balancing float.

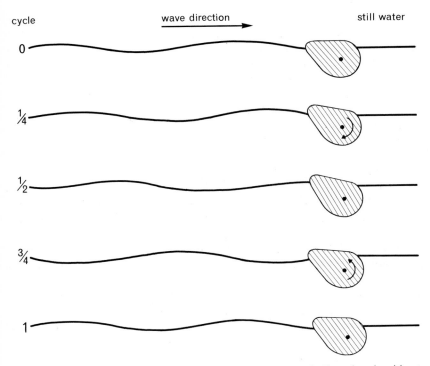

Figure 4.13 Action of Salter cam shape on waves, shown schematically and again without reactive balancing float.

oscillating water-column system, the "Cockerell contouring raft" and the "Russell rectifier" were to be studied.

In the first of these, devised by the inventor of the hovercraft, Sir Christopher Cockerell, the wave contour is followed by a series of hinged floats. Restriction in movement at the hinges, as between floats, is provided by power-abstraction devices. Efficiencies approaching 50% have been achieved. Under design wave conditions, there is again comparatively calm water in the lee of such a system. The Central Electricity Generating Board had already undertaken comparative tests of this and the "Salter duck" system at wave-tank as opposed to laboratory-flume scale, prior to the announcement of the more comprehensive programme of evaluation.

The "Russell rectifier", designed by R. C. H. Russell of the Hydraulics Research Station, is essentially a proposal to parallel the single-direction tidal-power scheme, but dealing with the comparatively short-period ocean waves rather than with the long-period waves of the tides. Water from wave peaks would be allowed into a higher-level chamber by a non-return door or gate system, and would pass into a lower chamber through a low-head turbine. The release would again be by non-return gates into trough conditions of the wave system. By its nature, such a system cannot hope to attain the theoretical efficiencies of the "Salter duck" units, but would have the advantage of depending to a much greater extent on existing technology and experience.

The ocean wave climate
The total gap between the laboratory situation of continuous regular unidirectional waves and realistic prototype conditions is a large one. Even during a short interval of time, there will be a spectrum of wave size and direction. However, it is considered that on good sites, such as exist west of the Hebrides, waves of suitable height and direction will be available—a situation arising from the great efficiency of transmission of large waves in the ocean. Theoretical and experimental studies have suggested that a Salter shape 50 m in basic diameter (which would correspond to about 90% maximum efficiency of absorption for ocean waves of 10 seconds period) would have efficiencies over 50% for periods between 7 and 17 seconds. However, current thought is towards a maximum diameter of 15 m for North Atlantic conditions. Efficiencies do not fall too quickly with small divergence from the normal presentation to the wave direction. Mooring arrangements would be made to allow adjustment of alignment, although the capacity for coverage of wide divergences is limited.

The problem conditions and their possible economic consequences
Fairly short periods of extreme conditions will inevitably occur, presenting differing hazards. When distant and local conditions combine to produce maximum storm waves, the whole system could be in danger. In practical

terms it seems inevitable that capacity to ride out the 100-year, or perhaps even the 500-year, storm would be a necessity for economic viability. This depends greatly on the economics of the system under normal conditions. The greater the economic breakthrough, the greater the risk that can be taken. Much development work will be necessary in the matter of storm conditions, once some idea of the economics of normal conditions emerges. Prototype tests will be necessary if the project reaches such a stage. However, much knowledge of storm circumstances can be obtained from model tests, given sufficient capital investment.

Thus energy-conversion operations would cease during the most adverse wave conditions—a situation which would also apply during periods of calm, which are infrequent in the sites proposed. Nevertheless, economic decisions would always have to be based on the production of energy, to be stored to a greater or lesser extent, rather than on the production of firm power. The need to store energy away from the prime converting system, and to cover periods of low or zero production, would have to be included in the assessment of economics.

Prospects for power from the waves

A development period of up to twenty years has been envisaged before large-scale production of energy could be practicable. During this time, full-scale tests would be undertaken into lengths of the wave-power-absorbing device sufficient to allow a complete assessment of the practicalities of the project. Even then, the overall economics would be imponderable because the economics of batch production would be difficult to forecast in such a new venture. In addition, although preliminary intermediate-scale tests may well establish a prototype shape and size quite close to the ideal, there are many possibilities in the application of the wave-derived torque. Salter first proposed that the prime energy conversion should be by a peripheral spline pump arrangement, based on a fluid-filled annulus, divided by longitudinal splines set axially in the basic circle of the Salter shape, as shown in figure 4.14. The presence of fluid in the annulus compartments between the inward-facing and outward-facing splines (with entry and egress controlled by an automatic pressure-sensitive-valve arrangement) would allow the resulting torque to be harnessed to produce electricity internally within the set of Salter-shape elements. Such a system presents development problems at least as complex as those presented in the development of rotating-element systems, with sufficient safeguards against adverse weather. If the matter is continued to full-size prototypes, then a significant commitment of resources will be involved before the final field tests are completed. There will inevitably be a diversity of possible

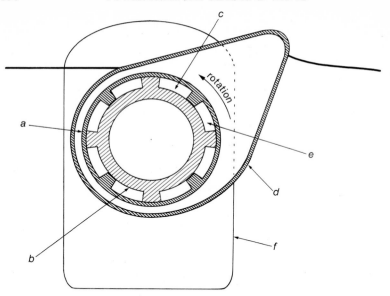

Figure 4.14 S. H. Salter's proposals for an integral spline pump arrangement in Salter-shape elements (after Salter).
(a) Fluid-containing cylinder moves with Salter shape. Inward-facing ridges with seal to inner cylinder.
(b) Inner-core cylinder with outward-facing ridges. Fixed by virtue of longitudinal connection over many Salter-shape elements.
(c) Working fluid being expressed from this compartment under high pressure, controlled by pressure-actuated non-return valves.
(d) Skin of Salter-shape element.
(e) Working fluid, re-cycled after energy production, flowing into the compartment as its volume increases, to be ready for reverse cycle of relative movement of cylinders.
(f) Vertical fin between elements of Salter shape.

total-energy linkages to be considered. Indeed it is unlikely that the best solution, which would emerge over a period of ten or more years from now, has yet been conceived. Salter and his team have already moved towards the adoption of more conventional torque-conversion linkages for the preliminary full-scale tests.

FURTHER READING

Bonnefille, R. (1976), "Les réalisations d'Electricité de France concernant l'énergie marémotrice," *La Houille Blanche*, No. 2, 1976.
Bernshtein, L. B. (1965), *Tidal Energy for Electric Power Plants*, Israel Program for Scientific Translations, Jerusalem.
Davey, N. (1923), *Studies in Tidal Power*, Constable, London.

Defant, A. (1961), *Physical Oceanography*, Pergamon, London.
Doodson, A. T. and Warburg, H. D. (1941), *Admiralty Manual of Tides*, H.M.S.O. London.
Gray, T. J. and Gashus, O. K. (editors) (1970), *Tidal Power* (Proceedings of an International Conference on the Utilisation of Tidal Power held May 24–29, 1970, at the Atlantic Industrial Research Institute, Nova Scotia Technical College, Halifax, Nova Scotia), 1972, Plenum, New York.
Reina, P. (1975), "The Magnificent Severn—Will it ever be tamed?" *New Civil Engineer*, No. 172, 11 Dec. pp. 26–27.
Richards, B. D. (1948), "Tidal Power; its Development and Utilisation," *Journal of the Institution of Civil Engineers*, **30**, 6, April, pp. 104–109.
Shaw, T. L. (1975), "Tidal Power and the Environment," *New Scientist*, October 23, pp. 202–6 (also T. L. Shaw (editor) (1975), *An Environmental Appraisal of the Severn Barrage*, Bristol University Press).
Swift-Hook, D. T. *et al.* (1975), "Characteristics of a Rocking Wave Power Device," *Nature*, **254**, April 10, pp. 504–6.
Wailes, R. (1941), "Tide Mills in England and Wales," *Journal of the Junior Institution of Engineers*, Pt. 4, No. 51, p. 91.
Wilson, E. M. (1973), "Energy from the Sea—Tidal Power," *Underwater Journal*, August, 175–86 (also "Alternative Energy Sources—Power from the Sea and its Economic Implications"—paper presented to the Canadian Electrical Association, Vancouver, B.C. Canada, in March, 1975).
Invited Authors (1974), "Energy Review," *Nature*, **249**, 5459, June 21 (including articles "Tidal Energy from the Severn Estuary" by T. L. Shaw and "Wave Power" by S. H. Salter).
Specialist Authors (1975), *C.E.G.B. Research*, No. 2, May. Special Issue on unconventional methods of transmission, storage and generation of power. Research Department of the Central Electricity Generating Board.

CHAPTER FIVE

DESALINATION

R. S. Silver and W. S. McCartney

Introduction

This chapter is devoted to desalination, in the sense of artificial production from seawater of fresh water for domestic and industrial purposes. Conventional water supply is based on the collection, storage and treatment of natural precipitation. This process actually includes the desalination of seawater, since rain or snow is produced from the sea by the energy of the sun. However, the term *desalination* is usually reserved for processes where the conversion is done artificially. As such, it constitutes—as do most of man's activities—an interference with the environment in which it takes place. Some aspects of this interference are beneficial and intended, while others may be unintended and perhaps adverse.

The circumstances in which a need for desalination arises and the interactions of this process with the environment will be discussed in this chapter.

The general features of the impact of desalination on the environment will be set out first. This philosophy leads to detailed consideration of the technological requirements. The actual technologies of desalination processes are then outlined. Finally, a brief summary of the relevant environmental situation is given.

Desalination and the environment
In man's inventions, desalination is unique in that its product is not a new commodity—like plastics and television—but is simply *an addition to the amount of something which is already available*. Fresh water is already available to a greater or lesser extent in all areas of the world, and can be

transported from areas with a surplus to those with a deficit. Such transport is familiar throughout civilized history—from the early irrigation channels along the Nile and Euphrates, and the aqueducts of the Romans, to the sophisticated water distribution systems of present-day Europe and America. It is doubtful whether any other single factor has such major environmental effects on a land area as does the availability or non-availability of fresh water. This determines not only plant and animal life, and hence agriculture, but all possibilities of industrial and social development (this aspect is examined in greater detail later).

There are parts of the world so arid over such large areas that the provision of fresh water in large quantities by transport from elsewhere is quite impracticable. There are even problems arising in non-arid parts of the world where the demands for fresh water are increasing rapidly, while the average climatic availability is either fixed or changing only very slowly. To the arid zones, desalination offers the basic essential for all development, and is a major factor in transforming their whole environment—agricultural, industrial, and social. To the developed non-arid countries whose demands are pressing against finite water resources (and therefore inevitably affecting their industrial and social environment) desalination offers a release of constraints.

This vital feature can be summarized by saying that every beneficial feature which fresh water has brought historically to the environment is precisely what desalination can now bring. Hence its impact on the environment is principally on *land* rather than on *sea*. In that sense, it could be argued that this chapter is out of place in the present volume.

The fundamental point, that the product of invention in desalination is not something new but an addition to something already existing, i.e. supplies of fresh water, is one which must be appreciated by all those whose consciousness of the risks of environmental damage makes them suspicious of new inventions. The product as such contains no environmental risk whatever. The interaction with the environment does not stop there, however, because any desalination process has as by-product a stream of brine more concentrated than the source water from which the product fresh water has been obtained. There is therefore a small interaction with the marine environment at the location of the plant, which is discussed in detail later in the chapter. But there is further major interaction with the land environment in the sense that any desalination process requires energy for its operation, and so in an energy-conscious world becomes a competitor in the demand for energy. Thus one essential aim in this discussion of desalination is to relate all the environmental advantages which it can provide via its fresh-water product to its demands on energy resources. These matters are considered in detail in the following section. Finally there are the less technical but nevertheless important environmental considerations, such as noise, amenity and aesthetic appearance—considerations which are relevant to any engineering installation. The position of desalination in these respects is included in the concluding survey.

For reference purposes in reading later parts of this chapter, it will be convenient now to say a little about the processes by which desalination is actually achieved. These processes are described in detail later, but the following preliminary remarks will be useful as background.

The only successfully established commercial process of desalination at present is that of distillation. This process involves the evaporation of water vapour from seawater, followed by its condensation into fresh water. Alternative technical possibilities exist, but have not yet reached commercial viability. These are electrodialysis, reverse osmosis, and freezing. The first operates by applying an electric potential across suitable membranes to provide ionic separation. The second is also a membrane process, but uses fluid pressure in place of electric potential. The third relies on the fact that when seawater is frozen the salt separates out between the ice crystals. Both electrodialysis and reverse osmosis have already proved viable for the purification of fluids with concentrations of solute much lower than seawater, and electrodialysis is fairly well established for brackish waters up to about 5000 ppm (seawater has approximately 35,000 ppm). It is known that the electrical energy consumption for electrodialysis of seawater will always be too high to be acceptable. Hence in the remainder of this chapter attention will be paid only to distillation, reverse osmosis, and freezing.

Water supply and desalination in an industrial society

Water, the forgotten industry

The production of a ton of steel requires over 40,000 gallons of water; of a ton of aluminium 300,000 gallons of water. Producing a ton of petrol in a refinery from crude oil needs 20,000 gallons of water, and a ton of artificial fibres from a chemical plant requires about 200,000 gallons of water. One gallon of beer produced uses about 350 gallons of water in course of its production. Considering these and similar statistics for other industries together, we find that over all the industrial activities of the modern world, one ton of an industrial product represents an average use of the order of 200 tons of water. In short, in terms of tonnage produced, the water supply industry of an industrial country constitutes about 99·5% of industrial activity, while all the rest constitutes only about 0·5%. That fact implies some very important economic fundamentals which have been too long ignored. The first is that industrial expansion or development is impossible without water in large quantities. The second is that unless the water can be obtained at a very much lower cost per unit than all the other materials involved in an industrial product, industrial production will be impractically expensive. Thus, for example, until recent years the price of water supplied in the United Kingdom was generally less than 1p per ton, corresponding to a contribution of only £2 per ton to the cost of an average industrial product. But now the cost of water in the UK is rising because of the rising costs of reservoirs and distribution schemes, and because the most convenient catchment areas were all fully exploited in the early period of industrial expansion. It was the fact that Europe had copious supplies of easily accessible water as well as of coal which made the Industrial Revolution possible. Without water, no such expansion could have taken place, despite the coal and despite the inventions on which the Industrial Revolution was based.

This aspect of economic history has not been fully realized by economic historians who, like most of us, have tended to take water for granted. But it is brought sharply into focus in the present day by the curious circumstance that so much of the world's oil fuel resource is found in the arid zones of the Middle East. The severe water deficiency there limits crucially the possibility of industrialization—or would have done so were it not for the development of desalination. The essential feature in the development of countries such as Kuwait and Saudi Arabia has been the use of a proportion of their oil resources to produce fresh water by desalination of seawater or of brackish water. The supplies of water thus gained provide the basis not only for domestic supplies but for the industrialization in which these countries are now participating.

To summarize this point we may say that in water-blessed countries such as the UK, the water industry, which in tonnage output dwarfs all other industries, has tended to be forgotten, while in water-deprived countries it has been recognized as the basic industry which has to be developed before any other.

Water and energy

While the western world has on the whole underrated the importance of its water, it is well known that the rate of development from the Industrial Revolution has been made possible only by the invention of the heat engine—to make things move by the burning of fuel instead of by human, animal, wind, or water, power. The dependence of all our industrial civilization on fuel energy is now well appreciated. One of the main ways in which we use fuel energy is in the generation of electrical power, which is essential for all modern industry and for modern domestic standards of living. Thus in the modern world, as industrialization and living standards develop, electricity consumption increases. We have already seen that water use also increases, and it is interesting to consider the ratio of water use to electricity use.

Evidently if we had assessed the ratio

$$\frac{\text{water use}}{\text{electricity use}}$$

in 1850, it would have been infinite, since electricity use was zero. In the period around 1930, in both USA and UK, the ratio was about 100 gallons of water consumption per kilowatt-hour of electricity consumption. In the next 30 years, although water consumption was increased, the use of electricity rose far more rapidly, and by 1960 the figure in the UK and USA was down to about 10 gallons per kilowatt-hour. This raises an interesting question: How low can this ratio become and still give an acceptable social standard? The problem has not been studied in great detail, but experience in the Middle East countries suggests that a minimum of 2 gallons per kilowatt-hour is necessary. If there is less water available than that in relation to electricity use in the community, conditions seem to be unacceptable. The UN Resources office has adopted a figure of 5 gallons per kilowatt-hour as an advised norm for planning purposes; i.e. if we expect a developing community to be using 500 MW electrical capacity (500,000 kilowatt-hours per hour), we must have a water supply of the order of 2,500,000 gallons per hour or 60 million gallons per day.

Against this norm of 5 gallons per kilowatt-hour advised for planning in developing countries, it is somewhat alarming to examine the most recent available UK figures. For the UK (excluding Northern Ireland) 1972 yearbook data give the figures $2·06 \times 10^{11}$ kilowatt-hours and $1·38 \times 10^{12}$ gallons—a ratio of 6·7 gallons per kilowatt-hour. This fall from 10 gal/kWh to 6·7 gal/kWh in 12 years indicates quite clearly that our industrial expansion is pressing on diminishing water resources. If this rate of fall continues, the UK will be down to the advised norm for planning in developing countries by 1978. It is a prospect which must call in question all the political and economic hopes of increasing growth rate to maintain the UK position as a leading industrial nation. Even if energy became more abundantly available at less cost, the water supply situation might be found to be a limitation.

It therefore becomes relevant to consider whether the UK will have to follow the example of the Middle East and introduce desalination to supplement its water supplies. The situation is less straightforward, for we do have substantial rainfall, and there are obvious alternatives to desalination. These alternatives taken together all imply a revolution in water management, and various aspects of this are now under study. These include new types of reservoirs, extensive development of systems by which contaminated water can be made suitable for re-use (perhaps several times), exploitation of underground water and of underground storage systems.

All such schemes are technically feasible, and no doubt many of them will be implemented. But one certain result is that water supply and water as a product are going to cost far more—even without inflation. An equally certain result is that these changes in water-supply procedure will impinge on many environmental aspects of life—on agriculture, on drainage, on canal and river transport, and even on local climate, as well as on many amenity and social aspects. For these reasons, the actual social cost will be exceedingly hard to assess beyond the flat statement that it will be much higher than the apparent cost. And when it is all done, it is still subject to rainfall variation. It is therefore quite likely that desalination (which gives freedom from rainfall vagaries, imposes minimum environmental reaction, and does not interfere with other water functions) may become attractive in certain parts of the UK. Until recently it would have been said that its acceptability or otherwise depended only on its comparative cost. Now, however, with the UK conscious of its energy crisis, the acceptability of desalination will depend not only on the comparative cost, but also on whether the cost is thought to represent a wise expenditure of energy.

This section has dealt with the general relations between energy and water use in an industrialized country, and the possible need for desalination is shown to arise from these considerations. We now must turn to consider what desalination itself demands in terms of energy.

Energy requirements of desalination

All possible processes of desalination require the consumption of energy. The energy may be consumed in two ways: as heat, or as work. While the latter form can, in principle, be obtained from wind or water power, the practical situation (of desalination being used to provide water independent of climatic conditions and in sufficient quantity) means that the work-consuming processes must be supplied with power generated from thermal power stations. Hence the energy quantity requirements of all desalination processes can be compared on the basis of their ultimate thermal energy needs. In this comparison it must be recalled that, for thermodynamic reasons, any thermal power station can convert only a portion of the thermal energy supply into work. Even in the best of modern practice, for each energy unit going in thermally, only 0·4 unit is produced as work, with 0·6 unit being rejected at a temperature lower than that of supply. A representative average of current practice may be taken as the division of each thermal energy unit supplied into a third of a unit of work output with two thirds remaining as lower-temperature heat output. Hence to assess the ultimate thermal-energy requirements of any work-consuming process, we shall multiply its work energy consumption by 3.

In making these assessments there are two figures which should be borne in mind. The first is that the thermodynamic potential-energy difference between seawater and fresh water is 3·53 kWh per 1000 gallons, giving a minimum theoretical thermal-energy requirement of 10·6 kWh per 1000 gallons, i.e. no possible process can ever be found which will use less thermal energy for desalination than this figure.

The second important figure to bear in mind is what is actually achieved at present in proved reliable desalination processes for large-scale water supply. The only proven reliable process established at the present time for supplying fresh water from seawater in quantities sufficient for population and industrial development is the multi-stage flash process of distillation (henceforth referred to as MSF). This process is one of the class which consumes its energy directly as heat, and the average requirement now established in practice is 315 kWh per 1000 gallons. Thus the best we can do reliably at present uses about 30 times the theoretical minimum. To the layman this may seem a very poor performance. Actually in terms of practical engineering it is very good. For technical reasons which dominate practice, many conversion processes are much further away from the thermodynamic energy minimum.

Against the background of these two figures, the minimum theoretical value and the representative figure of present-day reliable performance, we can now look at the possibilities of alternative processes. First we can make some assessment of the best we are likely to achieve in distillation as a result of research and development. It seems unlikely that MSF or any other viable distillation process will reach below 180 kWh per 1800 gallons.

Two other possible processes are by freezing and by reverse osmosis. The lowest expected *power* energy consumptions for these processes are respectively 35 kWh and 16 kWh per 1000 gallons. The corresponding *thermal* energy requirements for these power productions are 105 kWh and 48 kWh respectively. Thus, although neither of these processes has yet been proved viable nor reliable for desalination, we can picture the general comparison of thermal energy requirements in kWh per 1000 gallons shown in Table 5.1.

The figures in Table 5.1 show a very great energy advantage for reverse osmosis and also an apparently substantial advantage for freezing. However, before accepting these as conclusive, we have to consider the

DESALINATION

Table 5.1 Energy requirements of desalination per 1000 gallons (kWh)

	Distillation	Freezing	Reserve osmosis
Existing reliable established performance	315	—	—
Probable best attainable	180	105	48

practical situation of power production and distillation more closely. We have already noted that in power production, only part of thermal energy supplied is converted to power, while the remainder is rejected thermally at lower temperature. Now a point of crucial significance is that the *rejection at lower temperature is available for use in distillation*, so that the thermal energy requirements of the distillation process can be met by energy which would otherwise be rejected. This point is of very great importance and must be considered in some detail.

In a power station of normal design the rejection temperature is of order 33 °C. This temperature is too low to be of use in desalination. We must raise the rejection temperature to about 120 °C. This requires operating with higher back pressure. The result means that the power generating efficiency is reduced to about 75% of its previous value. Hence a station which on normal design produces $\frac{1}{3}$ power from 1 thermal unit, rejecting $\frac{2}{3}$ unit thermal at 33 °C, will produce $\frac{1}{4}$ unit power when modified, rejecting $\frac{3}{4}$ unit thermal at 120 °C. *The whole of this rejection at 120 °C can be used for thermal distillation.* The loss in power to obtain this is $\frac{1}{3} - \frac{1}{4} = \frac{1}{12}$. The amount of water obtainable by distillation from the thermal rejection now made available is, using the basis of 315 kWh required to produce 1000 gallons, $(1000/315) \times 0.75 = 2.4$ gallons. This is produced in association with $\frac{1}{4}$ unit of power, so that the combination station can satisfy a water/power use ratio of 9.6 gal/kWh. This is well above the recommended planning norm of 5 gal/kWh, and indeed above the use ratio in the UK. Evidently this facility is gained by the sacrifice of $\frac{1}{12}$ unit of power per $\frac{1}{3}$ unit of power which the station would have produced on normal design. Thus to meet the same power requirements as before, the energy consumption must be increased in the ratio $\frac{4}{3}$. The attainment entirely by desalination of a water supply sufficient to satisfy a use ratio of 9.6 gal/kWh therefore requires a 33% increase in energy consumption for a given power demand.

Lower use ratios can be satisfied without so much increase in energy consumption, since they can be met by only partial modification of the power station conditions. Suppose, for example, that we leave 50% of the power production at the original conditions while the other 50% is obtained, in association with water, at the modified conditions; we shall provide for a use ratio of 4·8 gal/kWh with an extra energy consumption of 16·7%. Table 5.2 shows the approximate extra energy consumption required for communities with varying water/power use ratios, where all the water supply is provided by distillation.

Table 5.2 Extra energy with distillation

Water/power use ratio (gallons/kWh)	2	4	5	6	8	9·6
Extra energy consumption %	6·9	13·9	17	20·8	27·7	33·3
Percentage of modification	20·8	41·6	52	62·5	83·3	100

When we consider the freezing and reverse-osmosis processes from the same point of view, the important fact is that these processes both require energy input in the form of power. Hence freezing, for example, requires 35 kWh of power input per 1000 gallons of water produced. If the use ratio is 5 gallons per kWh, this will represent an addition of 35 kWh to a power load of 200 kWh, i.e. an extra 17·5%. For a lower use ratio, the extra will be less, and for a higher use ratio, it will be more. Thus Tables 5.3 and 5.4 can be constructed for freezing and reverse osmosis, for comparison with Table 5.2 for distillation.

Table 5.3 Extra energy with freezing

Water/power use ratio (gallons/kWh)	2	4	5	6	8	10
Extra energy consumption %	7	14	17·5	21	28	35

Table 5.4 Extra energy with reverse osmosis

Water/power use ratio (gallons/kWh)	2	4	5	6	8	10
Extra energy consumption %	3·2	6·4	8	9·6	12·8	16

Several important deductions can be drawn from these tables. First it must be borne in mind that while the distillation figures are attainable already in established practice, the freezing and reverse-osmosis figures represent the probable best if ever these processes became practicable for water supply. They are not viable at present. The distillation figures can be improved by research and development. It is evident that the freezing process offers no advantage in energy consumption. It is also evident that no industrialized country dependent on imported energy can function if all its water supply is by distillation. In the UK, for example, with a present use ratio of 6·8 gal/kWh, the additional energy consumption would be about 23%. This shows up emphatically the value of our rainfall in providing the water resources which we do enjoy. Future pressure on these resources will undoubtedly drive the use ratio down. If it is to be prevented from falling below 5 gal/kWh, however, we shall certainly have to resort to some proportion of supply by desalination. A possible figure for this proportion for the UK is about 10%, and that would indicate an extra energy consumption of about 2%, which could be regarded as acceptable.

For the fuel-rich but water-deficient countries of the Middle East, the tables indicate clearly the energy sacrifice associated with necessary water development, and emphasize the value of water-control policies in their resource conservation.

These figures show dramatically the importance of desalination research and development directed to the possible reduction in the energy consumption of distillation and to attaining a practical reverse-osmosis process.

The cost of water by desalination

Any desalination installation for water supply has to be planned for the particular site. It is therefore impossible to give accurate total cost figures in a general discussion, because the optimal design of plant will depend upon local features, including local cost of fuel energy and of operational labour, the interest rates at which the capital can be obtained, and the intended life of the plant. However, some general guidance is obtainable from the energy costs. We have seen that the best reliable performance available at present is about 1000 gallons for a thermal energy input of 315 kWh. When this is obtained in conjunction with power production as discussed in the previous section, the actual required increase in thermal-energy consumption is about 120 kWh. Now the calorific value of fuel oil is about 5·4 kWh per lb. Thus the fuel requirement for 1000 gallons is about 22 lb, costing about 40p at present fuel prices in the UK. The capital and operating charges in an optionally designed plant will certainly be of the same order as the fuel charges, so that the cost of desalinated water in the UK at present (1975) must be in the region of £1 per 1000 gallons, i.e. 22p per ton. The days when water supply in the UK could be had for 1p per ton have vanished, and the present-day average figure is in the region of 10p per ton. Thus desalination for water supply in the UK at present will be directly economic only in areas where conventional supply costs are well above average.

However, if industrial growth is to continue at its present rate, the growing deficiency of water will increase costs very substantially, and desalination will undoubtedly become economic over a wider area. The crucial matter will be transferred from cost to energy itself, as outlined in the previous section.

Desalination and agriculture
While this section is entitled "Water Supply and Desalination in an Industrial Society", some reference must be made to agriculture, since without food supply no other industrial activity is possible. We have seen that the water-supply requirements of manufacturing industry are of the order of hundreds of tons per ton of product, averaging around 200 tons/ton. Interesting comparable statistics may be obtained by considering the water requirements per ton of agricultural crops. From the rainfall in the UK and the crop yields per acre, it can be estimated that the amount of water falling on a productive field is of the order of thousands of tons per ton of crop, averaging around 4000 tons per ton. Naturally it cannot be said that all this amount is actually necessary for the product. Yet when irrigation practice is examined, it seems that the requirement is still of the order of a thousand tons per ton. Hence it would probably be a safe generalization that the water requirements of agriculture per ton of crop are about 10 times those of manufacturing industry per ton of product.

The major part of agricultural industry worldwide relies entirely on direct precipitation on the fields and pays nothing for its water. With water requirements at the level indicated in the previous paragraph, it is quite obvious that agriculture costs soar enormously if the water has to be paid for. Even in countries where irrigation is necessary, it is doubtful whether a water cost of more than 5p per ton could be accepted, since this must contribute about £50 of cost per ton of raw crop. Recalling the figure of the order of £1 per ton for existing water desalination, it is obvious that there is no indication at present of desalination being economic for agriculture. Even the best predictions of reduced desalination costs from research or from development of reverse osmosis hold little hope of reaching a suitable level.

It is possible, however, that for high-value horticultural crops desalination will be beneficial. Indeed it has already been applied in a few special cases of this kind.

Methods of desalination

The distillation processes
The essential characteristic of any process of desalination by distillation is that it performs artificially the natural process upon which conventional water supply depends. The latter is obtained from precipitation of rain or snow. These in turn come from vapour which has been evaporated from the sea by the energy of the sun, condensed into rain or snow clouds in the atmosphere, carried by wind overland, and precipitated on the land, to form the rivers and lakes from which we collect our conventional

water. Desalination by distillation does the same, but we use the energy of fuel to vaporize the water from seawater, and condense it back again into liquid fresh water within the desalination apparatus. The conversion of water into vapour requires energy supply, and the condensation of that vapour back to liquid involves energy rejection. The amount required for evaporation and condensation is about 2·9 kWh for a gallon of liquid water. Hence a single boiling process in which we simply boiled off vapour and then condensed it again, releasing heat to ambient conditions, would require 2900 kWh per 1000 gallons. The actual present-day distillation performance of 315 kWh per 1000 gallons has therefore required something much more sophisticated than single boiling.

The fundamental notion by which this kind of improvement is obtained is that of doing the distillation in successive intervals. Thus, for example, taking a seawater temperature of 25 °C as ambient condition, a simple single boiling process would boil the seawater at about 102 °C and condense the vapour at near 25 °C. However, if we design a plant in which we maintain temperatures forming a series between the top and bottom values, we can accomplish the distillation in successive intervals, so that essentially the energy given out in condensing a portion of the vapour is used to provide that necessary to evaporate another portion.

This procedure of providing a series of temperature intervals is called either *multi-effect* or *multi-stage*, and the chamber in which a particular temperature is maintained is called an *effect* or a *stage*.

The energy given out in condensing in any of these intermediate chambers may be used either to boil off some more vapour, or it may be used to heat the brine liquid under pressure, so that boiling as such does not take place in the associated chamber. Nevertheless, when the pressure on the brine liquid is reduced, some of it flashes into vapour, so that the result of absorbing the energy while not allowing boiling is eventually the same, i.e. the release of a corresponding proportion of vapour. The process in which a plant is so designed that *all* vapour production is by flashing of pre-heated brine is called *flash distillation*. The process in which part of the vapour is produced by boiling is called *boiling distillation*, although in it always part of the vapour production—usually from 15% to 25%—is actually by flashing. It is usual to retain the term "effects" for the chambers in boiling distillation, and the term "stages" for those in flash distillation. Thus the respective processes with several intervals are referred to frequently as Multi-Effect Boiling (MEB), or Multi-Stage Flash, abbreviated to MSF.

It is a usual practice to refer to the *performance ratio*, or *gained output ratio*, of a distillation plant. This figure is the ratio of the energy required to evaporate unit mass of vapour from liquid at ambient conditions to the

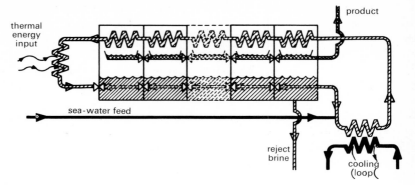

Figure 5.1 Simplified flow diagram of a multi-stage flash (MSF) plant for desalination of water.

energy actually required to produce unit mass of product from the plant. Denoting this performance ratio by the symbol R, the value which corresponds to the figure of 315 kWh per 1000 gallons which we have used is 2900 divided by 315, i.e. about 9·2. MSF plants realizing up to $R = 12$ have been successfully operated, and the figure of 9·2 chosen as typical for our purposes certainly represents what is viable and reliable at the present day in MSF operation.

MEB was the earlier form, as it developed naturally from the earliest single-boiling distillation which began on board steam ships a century ago. It reached its zenith in the mid-nineteen-fifties, achieving performance ratios of about 4 using six effects. But by 1960 it was virtually displaced by MSF. A simplified flow diagram of an MSF plant is shown in figure 5.1.

The crucial factors in the success of MSF were

(a) the operational simplicity arising from the entire elimination of boiling sections in the chambers, and
(b) the fact that the total temperature range can be divided into a greater number of smaller intervals than is possible for a given performance ratio with MEB, and so thermodynamic irreversibilities were reduced and efficiencies increased.

The basic patent for the new MSF process (R. S. Silver, 1957) specified that the number of stages should exceed twice the performance ratio. Thus performance ratios of 8 and upwards have been achieved using 20–40 stages.

Recently some efforts have been made to revive MEB in a rather different form. Previously the boiling sections took the form of pools of brine in which the heating tubes were immersed horizontally, higher-temperature condensation taking place within these tubes. Recent proposals provide for the boiling to occur in thin films of brine running down the internal or external surface of vertical tubes, while the higher-temperature condensation occurs on the external or internal surface respectively. The process is otherwise thermodynamically the same as the earlier form, but the film tube arrangement allows much better heat transfer than was possible in pool boiling. The virtue of MEB as compared with MSF in principle is that it requires substantially less *auxiliary* power consumption for pumps, although the main thermal-energy consumption does not achieve any improved value. There are, however, operational problems in maintaining the intended film flow upon which the expected heat transfer depends, and there are doubts as to whether this line of development will achieve the same level of reliability established by MSF.

There is one other form of distillation process of which mention should be made. It is the process of *vapour compression*. Its characteristic is that vapour which has been formed by boiling or flashing is compressed mechanically to a higher pressure and therefore to a higher saturation temperature. Its subsequent condensation can then deliver the rejected energy back to the brine to cause further evaporation. This process can be very useful for special small-capacity installations, but the problems of operation and maintenance of the mechanical compression system make it generally unsuitable for the large-capacity desalination plants needed for water supply.

Solar distillation uses the radiant energy of the sun directly. Up to the present it has consisted of shallow ponds of seawater contained within enclosures covered by glass or other suitable transparent material. The radiation of the sun causes evaporation from the seawater in the ponds and condensation occurs on the transparent covers. These are appropriately inclined for the condensate to flow down while still adhering to the surface. Appropriate collection devices then take off the resultant product. Such solar stills have been operated both as batch and continuous processes. They are attractive for tropical areas of high solar intensity where there is little availability of technical expertise. Their main disadvantage is that the area of land which must be covered by the ponds to provide sufficient product is substantial.

Until recently interest in solar distillation did not go beyond that kind of proposal. However, now that everyone is much more energy-conscious, it is probable that some of the more-sophisticated technological devices which are being worked out for the exploitation of solar energy will also be directed to desalination.

In an earlier section (p. 143) it was made clear that distillation is the only viable process at present. However, in the next two sections we shall outline for reference the alternative processes of reverse osmosis and freezing respectively. Then we shall return to the established distillation process of MSF and describe some of its subsidiary characteristics and their environmental consequences.

Reverse osmosis

When two systems have different values of Gibbs function (chemical potential) mass redistribution will occur, there being a net mass transfer from the system with the higher value of Gibbs function to that with the lower. This process will continue, if allowed to do so, until the systems have equality of Gibbs function. The Gibbs function of a solution is dependent on the pressure, temperature and concentration of solute in the solution. Under the same conditions of temperature and pressure, seawater has a lower chemical potential than pure water; and hence, if a sample of seawater is separated from a sample of pure water by a membrane permeable to water, the pure water will migrate through the membrane into the seawater, diluting it. This process can be reversed by increasing the pressure on the seawater side of the membrane to a value greater than the osmotic pressure, this pressure being a function of the concentration of non-volatile solute in the seawater. If the membrane is impermeable to salt, pure water will pass from the brine into the pure water on the other side of the membrane. This process is known as *reverse osmosis* and may be used to separate pure water from seawater, leaving a concentrated brine.

The difference between the Gibbs function for seawater and pure water is 3·5 kWh/1000 gal and for this difference an applied pressure of about 28 atmospheres is required. As the pure water is removed from the brine, the remaining seawater will become more concentrated in the region close to the membrane. The difference in free energies of the seawater and pure water will increase as the brine becomes more concentrated. To overcome this, either a sufficient flow of seawater must be maintained to minimize the increase in concentration, or the pressure applied to the seawater side of the membrane must be increased. In practice a combination of these two is employed, the applied pressure being of the order of 56 atmospheres, the work input required for this being about 21·42 kWh/1000 gal. The input energy requirement may be reduced by using the reject brine, which is at high pressure, to drive a hydraulic turbine, this turbine being used to supply some of the energy required for pressurizing the seawater. By this means the work input requirement may be reduced to about 16·4 kWh/1000 gal, corresponding to a thermal energy input of 49 kWh/1000 gal.

Most of the membranes used in reverse osmosis are of the order of

100 μm thick. In the design of reverse-osmosis plants, two physical characteristics play an important part. First, the membranes have little mechanical strength and are subjected to high pressure differentials across their faces. This means that the membranes require physical support while ensuring that the working surfaces remain in contact with both the pure water and seawater. The supports divide the membrane into small working areas capable of withstanding the applied pressure. There is still, however, danger of localized membrane failure, resulting in contamination of the product water. Secondly, the output of a reverse-osmosis plant is proportional to the area of working membrane surface in the plant. These two factors have resulted in most commercially available membranes and supports being produced in tubular form, as this type of arrangement provides a large active surface area in a small volume. At present most commercially available membranes are made of cellulose acetate and are supplied in the following basic formats.

The first of these consists of a perforated metal tube with the membrane located on the inside of the tube. The unpurified water flows through the tube, the purified water emerging through the perforations in the tube. A second arrangement is one in which a braided Terylene sleeve is mounted on a plastic rod. The Terylene is covered by the membrane. These membrane-covered rods are then mounted in parallel groups. The unpurified water flows in the spaces between the rods, the purified water passing through the membrane and Terylene, and being carried away in axial slots in the plastic rods. This type of arrangement is known as a "spaghetti module". Yet another arrangement uses a matrix of very fine hollow fibres, the active surface being on the outside of the fibres. The unpurified water passes over the outside of the fibres, the purified water being removed from the hollow centre of the fibre. So fine are these fibres that one million may be contained in a shell of diameter 1 cm giving an effective area of 12 m^2 in a length of 1 metre.

Membranes have been constructed from materials other than cellulose acetate, and much research is being done to evaluate and develop these new materials. The main lines of research in this field are, firstly, the development of high-flux membranes made of aromatic polyamides which have been shown capable of 98% rejection of chloride ions and, secondly, hollow-fibre membranes made of solid solutions of two glasses produced by means of temperature-controlled phase separation; but these have as yet been capable of only 88% salt rejection at a pressure of 125 atmospheres. They can, however, withstand pressures of up to 200 atmospheres.

At present levels of technological achievement there are two main problem areas with reverse osmosis. Firstly, the membranes are susceptible to fouling, both biological and non-biological, and because of this, pretreatment of the water is of prime importance. Because of the ion selectivity of the membrane, and the possibility of calcium sulphate from the seawater being deposited onto the membrane due to the localized high concentration at the membrane, the water must be softened using either Calgon (in large plants) or an ion-exchange resin (in small plants). The water must also be passed through a fine-mesh filter to prevent particles in the water larger than 10 μm from reaching the membrane and clogging the active surface. The presence of silica in the water may also affect the membrane. Secondly, the membranes currently available for reverse osmosis do not have infinite rejection capabilities, and some solute will be transferred from the solution to the pure solvent. By increasing the applied pressure, more pure solvent can be forced through the membrane, but as the characteristics of the membrane do not permit the passage of more solute, the effective salt rejection of the membrane is improved. This, however, has two disadvantages: (1) that increasing the applied pressure increases the required energy input, and (2) that increasing the applied pressure increases the likelihood of rupture of the membrane.

Although reverse osmosis is progressing in the field of desalination, there is as yet no commercially available "once-through" system. No membrane has as yet been developed which can sustain better than 99 per cent sodium chloride rejection and, as seawater contains on average about 35,000 ppm, 99 per cent rejection will not produce potable water comparable with that from a multi-stage flash plant.

Until membranes capable of providing a combination of high flux and high rejection become commercially available, water purification by reverse osmosis will be confined mainly to brackish water. Reverse osmosis is also used in the purification of industrial waste and sewage effluent, and in the concentrating of liquid feeds and liquid pharmaceutical products.

The freezing process

When seawater is cooled to a critical temperature, that temperature being dependent on the salt content of the water, salt-free ice crystals are formed in the water. These crystals may then be separated from the brine, washed to remove any residual salt traces, and allowed to melt in salt-free water. This is the basic technique employed in freeze desalination.

Let us consider two aspects of the process: thermal energy consumption and practical plant layout.

The brine used in the plant will initially be at a temperature in excess of that at which crystallization will occur. Thermal energy must therefore be transferred from it: (1) to reduce its temperature to freezing-point, and (2) to remove from it the latent heat of crystallization. The ice crystals thus formed must then be separated from the brine and have returned to them sufficient thermal energy to cause melting. The temperature at which crystallization begins in the brine is lower than that at which the ice crystals will melt when separated from the brine (due to the depressionary effect on freezing-point of a solvent of the addition of a non-volatile solute).

Energy is required to transfer the thermal load over the difference in temperatures. This ideal energy requirement is far below the minimum energy input attainable in practice. The energy transferred to whatever refrigerant is used in the plant must be later removed from it. Ordinary seawater is used as the heat sink but, as it will be at about 20 °C, the peak temperature in the refrigeration cycle must be higher than this to facilitate thermal-energy transfer in the required direction. Thus the thermal load must be transferred over a temperature range of perhaps 30 °C. If a simple cycle as described is used, the required energy input (thermal) may be of the order of 378 kWh/1000 gal.

The energy requirement can be drastically reduced by incorporating a secondary refrigeration unit. In this system a large amount of energy, mainly latent heat of crystallization, is pumped up a small temperature difference in the primary cycle, after which the latent heat is returned to the ice, leaving a small thermal load to be pumped up a large temperature difference in the secondary cycle. This means the work input to the plant may be reduced to about 44 kWh/1000 gal, this being equivalent (for modern compressors) to a thermal-energy input of 132 kWh/1000 gal.

A typical plant layout is shown in figure 5.2. The refrigerant most commonly used is butane. The seawater entering the plant is pre-cooled by the outgoing brine to reduce the thermal load on the primary cycle. The seawater then enters the crystallizer, where liquid butane is bubbled through the water, direct-contact heat exchange being used to minimize required temperature differences for thermal-energy transfer, and hence minimize the temperature range in the primary cycle. The butane vaporizes and consequently the seawater is cooled. Salt-free ice crystals are formed in the brine, which thus becomes more concentrated. This ice-brine mixture is then pumped to a wash colomn in which it rises, the ice crystals being compacted into a porous bed of ice. The brine flowing through the bed forces it to the top of the column, where it is washed with product water. The ice is removed to the melter, where it comes into contact with some of the vaporized butane from the crystallizer. The ice melts, and the mixture of water and condensed butane go to a decanter where the two liquids are separated. The product water then leaves the decanter. A small proportion of it is returned to the melter for washing the ice, while the remainder goes to the butane condenser and then to the

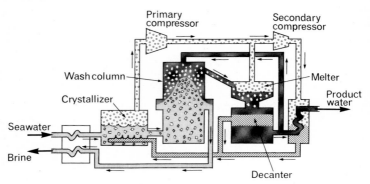

Figure 5.2 Freezing-process plant layout.

mains. The butane from the decanter is mixed with the remainder of the butane from the crystallizer, which has been compressed in the secondary compressor and condensed in the butane condenser. The liquid butane is then recycled to the crystallizer.

Distillation and power installation
In addition to the thermal-energy input required for MSF or MEB distillation, energy is also required for power to drive pumps. We have seen (p. 143) that the most economical use of thermal energy in these distillation systems involves steam which has already done work in turbine power production. Such an arrangement is frequently called a *dual-purpose* installation, since in a single station it produces both output power and water for sale. If, however, the power produced is so small that it is all absorbed in the distillation plant, we have only water for sale, and the result is what is called a *single-purpose* installation.

Most distillation plants throughout the world are dual-purpose installations. Often the distillation plant has been added to an already existing power station. Where this is done, only part of the main turbine steam flow is extracted to serve the distillation plant, and this proportion cannot usually exceed about 20%, since larger proportions would require appreciable modification to the low-pressure end of the turbine. Alternatively a dual-purpose installation may be specifically designed for both duties, in which case the desalination plant may replace the normal condenser entirely and accept the whole of the terminal steam flow. In such a case (see p. 142) the water/power production ratio can be about 10 gal/kWh. The power required by the distillation plant itself is only between 0·015 and 0·02 kWh per gallon. On this basis a dual-purpose installation producing 100 MW of total power can provide about one million gallons of water per hour (24 mgpd) and will absorb not more than 20 MW, leaving 80 MW for sale. More usually, however, as indicated earlier, only a part of the turbine steam will be used to supply heat to the desalination plant. The 20% rate would provide about 4·8 mgpd, using 4 MW for desalination power and leaving 96 MW for sale from the 100-MW capacity.

Correspondingly a single-purpose installation requires either auxiliary plant producing up to 0·02 kWh of power for every gallon of intended water product, or must buy in this amount of power. In most cases it is arranged to be self-supporting, i.e. produces its own power, frequently using direct steam-turbine drive for the main pumps instead of electric drive. Naturally, however, the major steam-raising load on the boiler of a single-purpose plant is not for the power production, but to provide the steam needed for the thermal-energy input to the distillation unit.

Finally a word should be said about *load matching*. In a single-purpose installation no problems of this nature arise, since it produces water only and can be put on or off load according to water demand alone. However, in a dual-purpose installation the situation can arise when at a certain period of the day electrical load may be small. If the water demand is simultaneously large, there may be insufficient steam flowing in the generating turbine to give the required heat to the distillation unit. This circumstance is conveniently met in practice by arranging for steam supply from the generating boiler to by-pass the turbine and go direct to the distillation unit. Conversely, if the electricity demand is high while water demand is low (or if the distillation unit is out of service for annual maintenance) the steam from the turbine cannot be absorbed in the distillation unit. Stand-by condenser capacity must be provided for the steam to meet such conditions. As is shown by the large number of dual-purpose installations now in service, such arrangements are very effective in practice.

Hence using the distillation process we can provide the necessary combination of power and water supply to match the needs of any community over a wide range of consumption ratio up to the infinite water/power ratio implied in single-purpose plant. Moreover it should be recalled that, unlike the electricity produced, the water produced can be stored to smooth out the effect of fluctuation on demand. It is usually found convenient to provide storage for a portion of the daily output of water.

Chemical aspects of distillation
Since the processes of evaporation and condensation, which are the basic essentials of distillation, are entirely physical, it might be thought that the chemistry of desalination plant would be unimportant. In fact, however, the practical operation of distillation plant is dominated by chemical aspects, even although the quantity of chemicals involved is very small in relation to the quantities of water processed in the plant. The reason for this praradox is that seawater contains not only sodium chloride, but also carbonates and sulphates which can produce scale deposition on the heat-transfer surfaces in the plant. Although the concentration of such substances is small (of the order of 0·01%), it means that quantities of the order of a ton of potentially scale-forming substances are taken into the plant for every million gallons of water product. If such scale were all allowed to deposit, the performance of the plant would dwindle within a few days. Hence, in all distillation operations, chemicals must be added to the seawater feed to prevent or control scale deposition.

Furthermore, the effectiveness of the possible chemical treatments depends on the temperature of operation. There are two main types of treatment: phosphate treatment, and acid treatment. The former can be used only up to about 100 °C and the latter up to about 120 °C. Higher-temperature operation is not possible because of calcium sulphate scale, against which no preventive treatment is known. The two methods mentioned above deal with carbonates only. These temperature limitations are important because, if we were free to choose higher temperatures, some reduction in the cost of desalination could be obtained. From this point of view, therefore, acid treatment is preferable. Nevertheless many designers and users prefer the lower-temperature phosphate treatment, because the acid treatment inevitably carries a greater risk of corrosion damage to the plant, especially if maloperation inadvertently occurs.

The scale problems also affect the energy consumption because satisfactory control of scale, even within the appropriate temperature limits, depends on the concentration to which the salts are allowed to build up in the plant. Normally we cannot allow the concentration to exceed twice that of normal seawater, which means that for every gallon of product fresh water we require a seawater input of two gallons, and there is a reject product of one gallon of concentrated brine. Although this concentration (7% as compared with normal seawater 3·5%) is not serious to normal marine life and is rapidly reduced in mixing when returned to the sea, the thermal energy carried in that brine has to be provided as part of the necessary input energy. In areas such as the Red Sea or Arabian Gulf, where the salt concentration of the input water is higher than that of normal seawater, the feed/product ratio may have to be increased from 2 to 3, so that the brine reject stream is twice the quantity of the product stream, and more seriously affects the energy consumption.

It follows that any seawater desalination plant must be regarded as having *two* products: the desired product, fresh water, and the unavoidable rejection product brine. The former is piped to wherever desired for supply, while the latter is returned to the sea. It is natural that from time to time inquiries are made as to whether this reject brine should not be used as a feed stock for production of sodium chloride, or chloride or related products like hydrochloric acid. However, since the reject brine is at most concentrated to only twice normal seawater it is still 93% pure water, and the roughly 7% concentration of NaCl is not sufficiently strong to be in general an economically viable feedstock. Nevertheless in a very few special cases it has been used as such.

The disposal of the rejected brine and its possible effects on the marine environment are discussed on page 160. The processes within the desalination plant are designed so that the fresh water produced is of very high purity, and all reaction products of chemical treatment are either carried away in the brine or discharged as gaseous diluents with the air driven out of the plant by the air ejectors. These air ejectors are necessary because much of the plant is under vacuum and inward leakage of air occurs, and because the intake seawater contains dissolved air which is released in the plant. The carbonates in seawater also break down in the plant, releasing gaseous CO_2 which must be expelled. The product fresh water from distillation is therefore almost entirely deaerated, and normally its total dissolved salt content does not exceed 50 parts per million. In contrast, most natural water supplies have upwards of 80 parts per million of dissolved solids, and for many industrial processes must be further treated by demineralization before. In such cases the product water from a distillation plant is valuable since it requires much less demineralization. Indeed, it is possible to design distillation plant which will produce water of a guaranteed purity, such that the solids content does not exceed one part per million, which is directly suitable for most industrial processes requiring high-purity water.

It is worth while remarking here that in this respect distillation differs very much from the freezing and reverse-osmosis processes which are not expected ever to be able to produce product waters with lower than 100 parts per million total solids.

As drinking water is tasteless and flat when totally devoid of dissolved air, it is usual to aerate the product water from a distillation plant when it is intended for domestic supply. Some authorities have also specified that calcium carbonate be *added* to the product water, in case of calcium deficiency in the diet.

Effects of desalination plant effluents on the marine environment
In an increasingly ecology-conscious society, any waste discharged into the oceans of the world must be investigated for its possible deletereous effect on mankind directly, and indirectly through damage to the marine environment. Fortunately, in a world intent on polluting the sea with undiluted sewage and industrial waste, the effluent from a distillation desalination plant is comparatively harmless, being in essence concentrated seawater. The effluent will also be warm, about 35 °C, and contain small traces of copper and some of the chemicals used in the prevention of scaling. The waste presents no toxicity problems, and therefore interest lies on its lesser effects on the marine fauna and flora.

We can dismiss any harmful effects on the macroscopic marine environment, as the brine outlet from the plant is located where tides will quickly disperse the effluent. This is a basic feature of the choice of location of the brine outlet, the other being to ensure that the outlet is positioned so that effluent cannot be carried by the tide to the seawater intake of the plant. As the total discharge from all the desalination plants in the world is only an infinitesimal proportion of the total volume of water in the world's ocean, it would appear that increasing use of desalination would have no harmful effects on the environment.

However, claiming that something causes no damage because it is quickly dispersed, still leaves the perhaps less-important question of how the marine environment close to the outlet from a desalination plant will be affected. Also, will it be affected in a beneficial or deleterious way?

The two most important aspects, quantitatively, of the effluent are salinity and temperature. The salinity factor affects both the fauna and flora in the sea. Algae and plankton are of particular interest, these being the least mobile geographically. They are more sensitive to low salinity than to high salinity, being able to withstand a salinity four times that of normal seawater, while the average effluent from a desalination plant has a salinity of only twice that of normal seawater. Although the survival rate of plankton and algae in the vicinity of the brine outlet may be unaffected, their growth rates will be affected, being in general at their maximum at normal seawater salinities (ignoring for the moment temperature effects). Although little work has been done on investigating this aspect, it is believed that exposure to brine of double normal salinity will produce a reduction in rate of growth by about 30 per cent.

The other equally important factor, temperature, has also an effect on the rate of growth of marine flora and fauna. Most aquatic organisms have no endogenous homeostatic mechanism for thermoregulation, i.e. they cannot maintain a steady body temperature independent of their environment and can only compensate for temperature change by metabolic adaptation. It has been shown experimentally that increase in temperature increases, slightly, the growth rates of most species. Unfortunately, as the high salinity of the effluent has a deletereous effect on growth, no use can be made of the warm effluent as, say, a growth medium for the production of "fruits de mer", such as molluscs and crustaceans.

Marine animals obtain the minerals required from the surrounding seawater via gills or epithelia. Some algae have in fact about 0·2 per cent by weight (dry) of copper, and most crustaceans will live happily on a diet of algae. Those species which require little copper for survival are unaffected by small excesses of it, and as the concentration of copper in desalination plant effluent (from the copper heat transfer surfaces) will be very small, it would appear that this would have little or no effect on the marine environment.

On the effect of the remnants of scale prevention additives, little or no work has been done but, again by virtue of the very low concentrations of these additives, it seems fair to assume that these have no effect on any important marine species.

The effluent so far described has been that which would be obtained from a multi-stage flash plant or a multiple-effect boiling plant. The effluent for the other main types of desalination processes, freezing or reverse osmosis, would be basically the same as that described, but at a lower temperature. This would remove the temperature factor for growth rate—producing, perhaps, a slight reduction in growth rate of marine flora and fauna.

From the foregoing arguments it would appear that, unlike most industrial and domestic waste which is dumped into the sea, the effect of effluent from desalination plants on the marine environment is minimal.

Appearance and amenity aspects of desalination plant
It is clear from the foregoing discussion that any process of desalination has to take in a stream of seawater feed and reject back to the sea a stream of brine, as well as provide its product stream. It follows that it is hardly practicable to site the plant far from the sea. The fresh-water product can be piped inland where required, but it would be unnecessarily expensive to take the larger-capacity seawater feed pipe far inland and have an equally large distance for brine return to the sea. Hence, in considering appearance and amenity aspects of desalination plant, it must be visualized as at or fairly near to the seashore. The general appearance will naturally

Figure 5.3 Single-purpose desalination plant on Jersey in the Channel Islands producing 1½ million gallons of fresh water a day.

Weir Group Ltd., Glasgow

depend on the type of process, while details of appearance will depend upon the particular design.

Thus distillation plant, which in its main bulk handles only steam, water, and brine streams, will tend in general to resemble power-station plant. Reverse osmosis handles only water and brine streams in its main bulk, and again will tend to resemble power plant. Freezing processes, however, have to handle refrigerant fluids as well as ice, water, and brine streams, and are necessarily more complicated in appearance, tending to resemble chemical plant. Within these generic resemblances, there can be considerable variety of detailed design. Our further discussion is limited to the known and established distillation process.

The appearance presented by the plant depends crucially on whether it is a single-purpose or dual-purpose installation (p. 156). In the latter, which is much the more frequent and likely, it will form only an addition or extension of plant on a site containing also all the familiar range of equipment associated with power generation (with the exception of cooling

DESALINATION 163

Figure 5.4 One of three seawater distillation plants producing a total of 4·4 million U.S. gallons of fresh water a day, built for the Malta Electricity Board and completed in 1970.
Weir Group Ltd., Glasgow

towers, which are not used at a seashore site). It is therefore unlikely that any "visual pollution" can be ascribed to distillation plant as such in a dual-purpose installation, since the "landscape design" will have to cater for the appearance of the whole complex. The humming noise inevitably associated with the operation of pumps in the distillation plant will also be a negligible addition to the usual power-station background noise level. Thus for the most frequent situation of distillation combined with power-station operation there is unlikely to be additional amenity objection.

When we turn to a single-purpose plant, the situation may be somewhat different. In this case the site will be principally occupied by the distillation plant, but must also provide a boiler and other auxiliary equipment for steam supply to the plant, which in the dual-purpose case is drawn from the power station. This whole single-purpose installation is then seen against whatever local shore background exists. Undoubtedly in some cases the result might be visually disturbing, and the noise output level

Figure 5.5 Three of five seawater distillation plants completed in 1966 for the Government of Kuwait. Each produces 1 million Imperial gallons of fresh water daily.

Weir Group Ltd., Glasgow

from the pumps will be noticeable. Its nature is, however, that of the steady low-pitched hum familiar in power generation, normally quite acceptable, and inaudible beyond the reasonable confines of the installation. The visual effect naturally depends on the site, the local terrain, and the design arrangement.

Figure 5.3 shows a single-purpose plant on Jersey in the Channel Islands (1971). The photograph is taken from the shore and shows the boiler house, complete with chimney stack, while the desalination plant itself is the smaller structure to the right of the picture. As indicated, the whole of the plant is sited in a disused quarry and only the chimney stack is visible from the countryside. Figures 5.4, 5.5 and 5.6 show various desalination plants installed in conjunction with power stations in Malta (1970), Kuwait (1966) and Abu Dhabi (1970). Unfortunately the power station complex as a whole is so large that it has not been possible to obtain photographs showing the whole of the power and water installations.

Figure 5.6 One of three seawater distillation plants, each producing 2 million gallons of fresh water daily, completed in Abu Dhabi in 1970 by Weir Westgarth Ltd. Three similar plants are on order.

Weir Group Ltd., Glasgow

Concluding survey

In this final section we very briefly survey desalination in relation to the environment in the light of the preceding discussions. The crucial factor is that desalination makes possible tremendous changes in the social environment of arid lands, provided that they are rich in energy. But it seems that the changes will be towards industrial development, to enhance manufactured exports which may, in turn, be used to purchase food—because the direct use of desalination for agriculture may be found to be too costly. This conclusion is rather saddening—implying that the desert is to bloom with manufactured goods rather than with crops. Such, at least, is the conclusion from the viewpoint of traditional economics.

However, the energy crisis in which we are now living is only one of the signs suggesting that traditional economics may have to be ruthlessly revised. The rate of sacrifice of all the world's resources on the altars of increasing productivity and international trade in manufactured goods has reached a stage where we must begin to question the reality of these goods. In the not-too-distant future, with an increasing population, it may appear absurd to produce manufactured goods for trading as an intermediary to obtaining food. A change in philosophy of this type could change relative values in a way which would direct desalination in energy-rich arid lands preferentially to agriculture.

For non-arid lands, the lesson of this survey is to value our existing water resources far more than we have yet done. The industrial growth rate to which we have been accustomed, and which politicians continually assume to be the answer to their employment problems, cannot continue without an increased use of water and consequent damage to all our environmental amenities. Desalination offers an alleviation which can help to maintain both amenity and industrial growth—but only if we have access to sufficient energy. Happy the land that has both energy and fresh water!

FURTHER READING

Clawson, Marion and Landsberg, H. H. (editors) (1972), *Desalting Seawater*, Gordon & Breach, London.
Kinne, O. (editor) (1975), *Marine Ecology*, Wiley-Interscience, London.
Newell, G. E. and R. C. (1963), *Marine Plankton*, Hutchinson Educational, London.
Porteous, A. (1975), *Saline Water Distillation Processes*, Longman, London.
Sourirajan, S. (1970), *Reverse Osmosis*, Logos Press, London.
Spiegler, K. S. (1966), *Principles of Desalination*, Academic Press, London.

Index

Abu Dhabi desalination plant 165
airborne oil particles, fall-out 50
algae 3, 12, 15
   and eutrophication 97
   green 98
   salinity effect 160
aluminium manufacture, water demand 140
anchoveta 26, 36, 39, 40
animal behaviour, effects of oil 58
Antarctic Convergence 16, 40
aquaculture 41
Arenicola 85, 87, 89, 90
arrow-worms 27

bacteria and breakdown of oil 52
   in sea 8
ballast, discharge 47
   pollution hazard 47
Baltic herring 31
barrage construction 119
Bay of Fundy 127
beer manufacture, water demand 140
Bélidor 104
benthic animals 26
   organisms 4
benthos 20
Bernshtein, L. B. 103
brackish water, desalination 140
bulb turbines 112

Calanus 20, 21, 22
Calgon 154
carbon fixation 13
carcinogenesis with oil 63
carcinogens from oil not in food chains 65
carnivores 6
cavitation prevention 111
chemical aspects of distillation 158
   pollutants, coastal 85
      sources 72, 73
china clay waste, 80, 81
chlorine toxicity 75
chlorinity 94
chlorophyll 10, 12, 14, 19

Clean Seas Code 48
coal bings 91
   pollution 84
Cockerell, Sir Christopher 133
cod 8, 40
compensation depth 11, 14
concentration of chemicals by oil 59
continental shelf 4, 16, 26, 70
Cook Inlet 127
copepods 24
cost of water by desalination 147
crude oil components 50, 51
crustaceans 8

Davey, Norman 104, 128
decomposers 4, 24
Defant, A. 109
demersal fish 8
desalination 138
   and agriculture 148
   and horticulture 148
   assessment of UK need 142
   by freezing 144
   by reverse osmosis 152
   chemical aspects 158
   economic aspects 165
      UK 147
   effect on the marine environment 160
   energy cost 143
   integration with power production 145, 146, 156, 157
   limit of purity of water produced 159
   methods 148
   plant, appearance and amenity 161
diatoms 3
dinoflagellates 3
distillation by vapour compression 151
   chemical aspects 158
   limit of purity of water produced 159
   plant performance ratio 149
   plant gained output ratio 149
   process 148
diurnal inequality 108
Dogger Bank 41

dumping 91
  in the sea 75, 77

ecological efficiency 21–24, 33
economics of desalination 165
  desalination in UK 147
  power generation 117
  water production 141
efficiency of electricity generation 143
efficient prey 31, 32
electricity generation, efficiency 143
  from tidal power 110
electrodialysis 140
energy cost of desalination 143, 144, 145
equilibrium tides 106
euphotic zone 11, 14–16
eutrophication, 71, 74, 96

Fentzloff, H. E. 128
fish as source of protein 1
  bottom-living 8
  catch per unit effort 32
  deep-water 40
  demersal 8
  eggs, effect of oil 58
  farming 1, 36, 40, 42–45
  global yield 1, 24, 25
  meal 40, 42
  ponds 42, 44
  potential harvest 26
  production 21
  recruitment 41
fisheries, condition for maximum yield 32
  factors influencing yield 36
  future 39
  global yield 40, 44
  growth of yield 26
  model 34
  overfishing 37
  Peruvian 26
  prediction of yield 38
  problems 37
  regulation 34, 36, 37, 45
  regulations international efforts 38
  sites 70
  unconventional 40
fishing effort, effect on catch 33
fixation of carbon 10
flagellates 3
food chains 6–8, 20, 22–24
  effect of oil 59
food webs 31
foods, global production 2
freezing, desalination by 140, 154, 155

Gibbs function 152
global production 17
green algae 98
growth efficiency 20
Gulf Stream 4
gypsum concentration in seawater 94
  spoil heaps 92, 93

Hanseatic League 31
herbivores 6, 24
  zooplanktonic 22
herring 6, 22–24, 32, 35, 38–40
  Baltic 31
  development 27
  North Sea 28
  place in food chain 29
hydrocarbons from plants 50
hydro-electric schemes 112

Iceland—Faroes Ridge 16
illumination, influence on photosynthesis 10
Industrial Revolution 103, 141
inorganic chemicals, effects on aquatic
  organisms 76, 77
  wastes 70
ion-exchange resin 154
iron waste deposits 78
irrigation 148

jellyfish 5, 27
Jersey desalination plant 162

Kaplan turbine for tidal barrage 111
kerosine 46
Kislaya Guba 118, 125
krill 40
Kuwait desalination plant 164

La Rance, tidal-power installation 104–6, 118
  122, 124, 129
  turbines 113
land fill, environmental risks 100
larvae, effect of oil 58
limpets 57
  effect of oil 61
Littorina 56
load matching 157
lubricating oil 50
lugworm (*Arenicola*) 85, 87, 89, 90

Malta desalination plant 163
man as predator of fish 2
mariculture 40
MEB (multi-effect boiling) distillation 149,
  150, 151, 156, 161

INDEX

Mevagissey Bay 80, 81
Middle East, importance of water-control policies 147
    water 141
migration of zooplankton 4
Milford Haven 56, 61, 62
milkfish 44
molluscs, effect of oil on 55
Monodonta 55, 56
MSF (multi-stage flash) distillation 144, 149–151, 156, 161
mussels 43

NCB (National Coal Board) 84
Nereis 78, 79
North Sea fisheries 28, 29, 35, 38
Nucella 55
nutrient cycling 6, 8, 13, 31
    regeneration 14, 16

ocean wave climate 134
offshore oil production 47
oil, acute toxicity 54
    components, degradation of 52
    consumption 46
    crude, components 50
    industry, development 46
    lethal concentration 54
    limit of detection in drinking water 60
    lubricating 50
    pollution and cancer 63
    refinery effluent, ecological consequences 62
    slicks, dispersal 51
oil spillage 46, 47
    biological effects 46, 52
    containment 66, 67
    ecological consequences 57, 60
    effect on amenity 63
        fish 60
        limpets 61
        marine plants 53
        marine mammals 52
        oysters 60
        salt marsh 63, 64
        seaweed 61
    risks in treatment by detergents 68
    sources 49
    sublethal effects 58
    toxic effects 53
    toxicity test results 56
    treatment 65
oil, spilt, properties 50
osmosis, reverse, 140
overfishing 37, 39

oysters 43
    effect of oil 60

paraffin 46
Passamaquoddy Bay 127
pelagic food chains 22
    organisms 4
Peruvian fisheries 36
Petroleum and the Continental Shelf of NW Europe 62
pH and toxicity 74
    variations 73
phosphogypsum 92–96
photosynthesis 6, 10, 11, 15
    at various depths 11
    conversion efficiency 18, 19
    influence of illumination and temperature 10
phytoplankton 3, 4, 8, 9, 12–14, 16, 19–22, 39, 42, 43, 73, 97
    increase in spring 15
plaice 8, 29, 30, 34, 38, 41, 43
plankton 3–5, 26, 56, 58
    effect of salinity 160
pollution 30, 43
    from coal bings 91
    oil marine 46
polyculture 44
power from waves 130
    station efficiency 143
predators, prudent 31
prey efficiency 31
primary production 9, 15, 22, 25, 32, 44
    efficiency 18
    growth efficiency 29
    seasonal and regional variations 14
production at higher trophic levels 20
protein shortage 1
pumped-storage link with tidal power 123

recruitment 33
    of fish 42
red tides 97, 98
respiration 6
reverse osmosis 144, 152, 154
    technical difficulties 153
Russell, R. C. H. 134

salinity 94
    factor 160
salmon 42, 43
salmonid farming 42
Salter, S. H. 130, 135, 136
    cam shapes 133

saltmarshes 99
  and red tides 100
  effect of oil spills 63, 64
San José Gulf 128
sandhopper 7
scale prevention in distillation plant 158
seabirds 8
  effects of oil on 52
sea gooseberries 27
sea-trout 42, 43
seawater, carbon dioxide in 72
  desalination by freezing 154, 155
    reverse osmosis 152
  pH 72
seaweed 4, 7
seine net 27
sensitivity to oil 55
Severn Estuary barrage 125, 126
sewage sludge deposits 78
  treatment 98
slag disposal 82
slagcrete 84
sluices 117
Smeaton, John 103
solar distillation 151
  ponds 151
Solway Firth 71, 82, 83
Spartina 64
spillage, oil 47
spilt oil properties 50
standing crop 9, 14, 16, 19, 21, 32
  terrestrial and marine plants 12
steel making, water demands 140
Studies in Tidal Power 104
Study of Critical Environmental Problems 49
syzygy 107, 108

tambaks 44
temperature, influence on photosynthesis 10
territorial waters 38
Thailand 26
Thames estuary pollution 97
thermocline 14, 15
tidal basins 102, 103
  conditions for a potential power project 108
tidal power, basin arrangements 122
  economic assessment 128
  installation, integration with grid 120
  operation methods 121
  potential output 129

tidal predictions 108
  range 107
tide mills 102, 104
tide-producing forces 107
tides, equilibrium 106
Torrey Canyon 56, 60, 68
toxicity of various oils 57
trace elements and *Nereis* 78, 79
trace metals, availability v concentration 91
  in coastal pollution 85
  in shore sediments 86, 87
trawling 27
trophic levels 6, 22, 24
tuna 26, 39
turbine, bulb 112
  designs for tidal power 114
  developments 112
  Pelton-wheel 117
  S-tube 115
  straight-flow 115

upwelling 4, 14, 16, 25

vapour compression, distillation by 151
"Vikoma" seapack system 66, 67
visual pollution by desalination plant 163
VLCC (very large crude carriers) 47

wastes, inorganic 70
water and energy 141
  cost of, and effect on industrial development 141
  mills 102
  production 141
Water Quality Criteria 75
water required by industry 140
  resources and standard of living 142
  supply, cost of increasing 143
    industrial demands 140
  wheels 102, 103, 104
wave action, power available 131
Wave Energy Steering Committee 132
wave power 130
  economic aspects 135
  prospects 135

year class 34

zooplankton 4, 5, 8, 9, 13–16, 19–23, 27, 39, 42

Kingston College of Further Education
Books are to be returned on or before the last date stamped below.

10. JAN. 1983	30 NOV 1987	18 MAR 2002
-9 FEB 1983	29 JAN 1988	12 MAY 2005
-x. FEB 1983	20 OCT 1989	19 JAN 2017
-5 MAY 1983		
-1. FEB 1985	30 OCT 1995	
-6. JUN. 1985	MS	
16. OCT. 1985	28 FEB 1997	
16. OCT. 1985	12 MAY 1997	
24. JUN. 1986	19 MAR 1998	
22. OCT. 1986	10. DEC 2001	
7 OCT 1987		SC10243
21 OCT 1987		

32087

00024525

ENVIRONMENT AND MAN: VOLUME FIVE

Titles in this Series

Volume 1   *Energy Resources and the Environment*

Volume 2   *Food, Agriculture and the Environment*

Volume 3   *Health and the Environment*

Volume 4   *Reclamation*

Volume 5   *The Marine Environment*